D0148757

*Curves and singularities*

'Singularity is almost invariably a clue'

# Curves and singularities

*A geometrical introduction to singularity theory*

## J. W. BRUCE

*Department of Pure Mathematics, University of Newcastle upon Tyne*

## P. J. GIBLIN

*Department of Pure Mathematics, University of Liverpool*

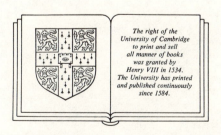

The right of the
University of Cambridge
to print and sell
all manner of books
was granted by
Henry VIII in 1534.
The University has printed
and published continuously
since 1584.

## CAMBRIDGE UNIVERSITY PRESS

*Cambridge*
*New York New Rochelle*
*Melbourne Sydney*

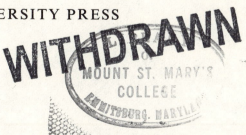

WITHDRAWN

MOUNT ST. MARY'S
COLLEGE
EMMITSBURG, MARYLAND

Published by the Press Syndicate of the University of Cambridge
The Pitt Building, Trumpington Street, Cambridge CB2 1RP
32 East 57th Street, New York, NY 10022, USA
10 Stamford Road, Oakleigh, Melbourne 3166, Australia

© Cambridge University Press 1984

First published 1984

Reprinted with corrections 1987

Printed in Great Britain at the University Press, Cambridge

Library of Congress catalogue card number: 83-14456

*British Library cataloguing in publication data*

Bruce, J. W.
Curves and singularities.
1. Geometry, Differential
I. Title II. Giblin, P. J.
516.3′6   QA641

ISBN 0 521 24945 7 hard covers
ISBN 0 521 27091 X paperback

DE

For Linda and Rachel

# Contents

# Preface

'I don't think you need alarm yourself,' said I.
'I have usually found that there is method in his
madness.' 'Some folk might say there was madness in
his method,' muttered the Inspector.
(*The Memoirs of Sherlock Holmes*)

The object of this book is to introduce to a new generation of students an area of mathematics that has received a tremendous impetus during the last twenty years or so from developments in singularity theory.

The differential geometry of curves, families of curves and surfaces in Euclidean space has fascinated mathematicians and users of mathematics since Newton's time. A minor revolution in mathematical thought and technique occurred during the 1960s, largely through the inventive genius of the French mathematician René Thom. His ideas (partly inspired by the earlier researches of H. Whitney) gave birth to what is now called singularity theory, a term which includes catastrophes and bifurcations. Not only has singularity theory made precise sense of what many of the earlier writers on differential geometry were groping to say (as so often happens, their instinct was uncannily good but they lacked the proper formal setting for their ideas); it has also made possible a richness of detail that would have stirred the imagination of any of the great geometers of the past.

Thom applied his ideas to many fields besides geometry, for example in his famous (but very difficult) book (Thom, 1975). Since then he and others have sharpened and modified the ideas so that the applications to science, potential and actual, are now very impressive – also occasionally very controversial. We believe that the best way to introduce singularity theory is to show it in action in a situation where there is no doubt about what is going on and which requires a minimum of specialist knowledge. For that reason we have included only geometrical applications here, apart from a brief look at catastrophe machines in chapter 1, and some geometric optics and differential equations in chapter 7. Further, in order to keep the technicalities under control, we have restricted ourselves to geometrical applications of singularity theory that hinge on the concepts of transversality and of unfolding a smooth function of one variable. Thus we include the geometry of plane and space curves, envelopes of one-parameter families of curves and surfaces, evolutes, duals, contact, apparent contours of surfaces, caustics, symmetry sets, maps from the plane to the plane and the 'generic geometry' of plane and space curves – properties which hold for 'almost all' curves. (Even if these topics do not mean much to you, we hope that a casual glance through the diagrams will convince you that they are potentially interesting!)

We do not assume previous knowledge of curves or surfaces; in chapter 2 we start the study of curves from scratch, biassing our account very much in favour of singularities of real-valued functions on the curves and contact with lines, circles, planes and spheres.

The book splits, we hope not literally, into two parts. The first seven chapters are probably enough for a final year (or, in some places, penultimate year) undergraduate course of 30–35 lectures. Chapter 11 could also give an informal finale to such a course. Smooth curves in the plane and in space, contact, envelopes, smooth parametrized manifolds and unfoldings are studied in detail, and many applications and illustrations of these ideas are given both in passing and more systematically in chapter 7. (We kept thinking of more applications, but had to call a halt somewhere.) We hope that a course of this kind might provide a useful antidote to the currently fashionable packaged and axiomatic courses at the undergraduate level, and might also bring back some real geometry to degree courses. (Is it unreasonable to believe that a thorough grasp of something rather concrete and not too difficult can be better than a vague understanding of high-faluting technical mathematics?) The only prerequisites are some knowledge of linear algebra (matrices and linear maps) and several-variable calculus. (We state and use, but do not prove, the Inverse Function Theorem.) One comparatively deep and difficult result, which gives the existence of 'universal unfoldings', is discussed in some detail, but not proved, in chapter 6.

We regard chapters 8–10 as somewhat more sophisticated than chapters 1–7; they are suitable for relatively advanced undergraduates or first year postgraduates. The pace is rather faster and we assume a good understanding of what has gone before as well as some familiarity with ideas such as compactness. Also in chapter 9 not all the details of all proofs are given. We cover transversality and the application of this crucial idea, via Thom's transversality lemma, to the proofs of results about 'general' curves – properties which hold for 'almost all' curves. In chapter 10 we give a proof of the 'universal unfolding' theorem used in chapter 6, not in the full generality assumed there (such is neither desirable nor feasible in a course at this level), but under the assumption that our functions, curves and so on are *analytic* rather than just smooth. Thus our treatment is complete in this case, which is quite sufficiently interesting in its own right. The proof given in chapter 10 is not the 'standard' one via the Preparation Theorem, but instead is an adaptation of a more general argument of Kas and Schlessinger. We believe it shows rather clearly why the criterion for a versal unfolding has the form that it does. The first part of chapter 10 is not at all technical. Finally, in chapter 11, we briefly survey the prospect from the vantage point gained so far.

We have included a large number of exercises, many with hints for solution. We believe that mathematics can only be truly said to be going on when the mathematician (professional or student, but we are all students of a kind) is solving problems, and one of the attractions of this subject matter is that it lends itself very well and at every stage to the solution of explicit problems. We also hope that the inclusion of numerous exercises will encourage non-experts to try teaching this concrete and (we hope) enjoyable material, much of which has not appeared before in book form.

The results of some exercises are used later in the book; these are always provided with adequate hints, so can be regarded as small extensions of the text with some details missing. The sections in small print, on the other hand, are peripheral to the main issues and can be skipped without loss of connexion – though with some loss of entertainment, for there is at least one murky backwater of the subject represented there.

Most of the material in this book has been taught in a final year undergraduate course at Liverpool University or in a first year postgraduate course at University College Cork (or in both). We should like to thank our colleagues at these places, and of course our students, for their help, unwitting or otherwise, in getting the material straight. In particular we thank Professor C. T. C. Wall, of Liverpool, for getting us interested in this subject in the first place. The second author spent much of the writing period as Visiting Professor or Visiting Scholar at the University of North Carolina at Chapel Hill and the University of California at Berkeley. He thanks these institutions for their hospitality, the first of them, and Liverpool University, for financial support, and the faculty members, particularly Professor J. N. Damon of UNC, for support of a moral and mathematical kind. The computer pictures were produced at Liverpool, using an IBM 4341 computer and a CalComp incremental plotter. The mottoes for chapters and elsewhere all come from the exploits of Sherlock Holmes, in the stories of Sir Arthur Conan Doyle. The frontispiece was drawn for us by Heather Harrison.

# 1

## *Introductory example: a gravitational catastrophe machine*

'I am afraid that I rather give myself away when I
explain,' said he. 'Results without causes are much
more impressive.' (*The Stockbroker's Clerk*)

It is well established that one should never begin a talk – or presumably a book – with an apology. We apologize, therefore, for apologizing that despite the title of this chapter our book is *not* primarily about catastrophe theory. The reason for our beginning with a gravitational catastrophe machine is that it exemplifies, in a vivid way, many of the ideas we shall study in detail later, such as functions on a curve, envelopes, surfaces, projections, evolutes and bifurcation sets. These ideas are merely touched on in the present chapter: do not expect to understand all the details yet.

The gravitational catastrophe machine was invented by T. Poston and is discussed in the well-known book on the subject (Poston and Stewart, 1978). Other introductions to catastrophe theory can be found in Zeeman (1977), Poston and Stewart (1976), Saunders (1980).

Consider a parabola, cut off by a line (perpendicular to the axis say), as in fig. 1.1. Imagine the region enclosed to be a lamina (thin sheet) that is

**Fig. 1.1.** Parabolic lamina in a vertical plane

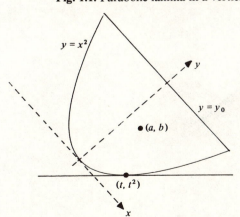

1

constrained to move in a vertical plane, resting on a horizontal line; we seek the position of stable equilibrium. We do not assume the lamina to be of uniform density; in fact let its centre of gravity be at the point $(a, b)$ referred to axes $x$ and $y$ as shown, relative to which the equation of the parabola is $y = x^2$. (The point $(a, b)$ is often said to be in the *control space*: the values of $a$ and $b$ control the behaviour of the lamina.) Then we are looking for positions that give (local or global) minima of the potential energy of the lamina. The latter is easy to calculate: it is $mgh$, where $m$ is the mass, $g$ the acceleration due to gravity and $h$ the height of the centre of gravity above some base line, which we take to be the line on which the lamina rests. Thus, if the parabola makes contact with the line at $(t, t^2)$, then $h$ is the distance from $(a, b)$ to the tangent $y - 2tx + t^2 = 0$. The poten-

**Fig. 1.2.** The surface of normals to a parabola

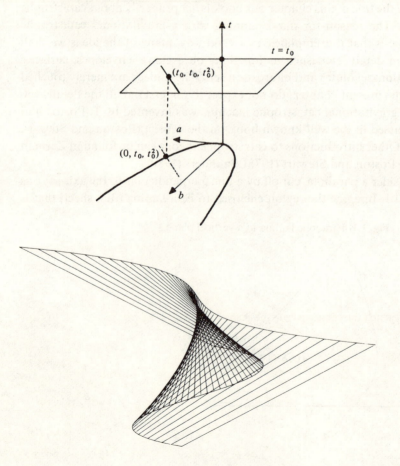

tial energy is therefore

**1.1** $$V(t) = \frac{b - 2ta + t^2}{(1 + 4t^2)^{\frac{1}{2}}} mg,$$

and the mathematical problem is simply, for given $a$ and $b$, to find the value or values of $t$ which minimize $V(t)$. We find that

**1.2** $$\frac{(1 + 4t^2)^{\frac{3}{2}}}{2mg} V'(t) = U(t), \text{ where } U(t) = 2t^3 + t(1 - 2b) - a,$$

so that we want $U(t) = 0$ for a turning point and $U$ to change from negative to positive at this $t$ for a minimum. Notice that $U$ depends on the given $a$ and $b$; as $a$ and $b$ *vary* we obtain a *two-parameter family* of functions. Such families will occupy us a good deal later on.

Now $2t^3 + t(1 - 2y) - x = 0$ is precisely the equation of the *normal* to the parabola at $(t, t^2)$. Thus we obtain first the unsurprising fact that for equilibrium $(a, b)$ must be vertically above the point of contact $(t, t^2)$, for the normal at $(t, t^2)$ is the vertical line through that point. We shall return to the question of minima shortly.

It is instructive to regard $2t^3 + t(1 - 2b) - a = 0$ as defining a *surface M* (called the *catastrophe surface* of $V$) in the three-dimensional space with coordinates $(t, a, b)$. With the $t$-axis vertical, $t = t_0$ is a *horizontal* plane which meets $M$ in the line $t - t_0 = 2t_0^3 + t_0(1 - 2b) - a = 0$. This is a normal line to the parabola in the plane $t = t_0$ given by the equation $b = a^2$, the normal being at the point $t = t_0$, $a = t_0$, $b = t_0^2$. What this means geometrically is that $M$ is obtained by taking the normals to the parabola $b = a^2$ in the $(a, b)$-plane and moving the normal at $(t_0, t_0^2)$ vertically to the height $t_0$. The normals are spread out to form a surface by using the $t$-direction* (fig. 1.2).

The *vertical* plane $b = b_0$ meets $M$ in the cubic curve $2t^3 + t(1 - 2b_0) - a = 0$. As $b$ increases through $b_0 = \frac{1}{2}$ this curve acquires a maximum and a minimum, thinking of $a$ as a function of $t$.

**1.3     Exercise**

Sketch the cubic curve in the plane $b = b_0$ for $b_0 = 0, \frac{1}{2}$ and $\frac{3}{2}$.

---

* It is possible to make a model of the surface $M$ using cotton threads for these lines, passing through holes drilled in perspex (or wooden) sides $b = \text{const}$. Here, in centimetre units, are good dimensions to use. Try the curve $32b = a^2$ with planes $b = 0, b = 24$, taking $t = 0, \pm 0.8, \pm 1.6, \ldots, \pm 16$. The face $b = 0$ can be about 36 in $t$-direction $\times 24$ in $a$-direction $((t, a, b) = (0, 0, 0)$ at the centre) and the face $b = 24$ can be about 36 in $t$-direction $\times 12$ in $a$-direction $((0, 0, 24)$ at the centre).

We draw the folded surface $M$ as in fig. 1.3.

For given $a$ and $b$ the vertical line through $(0, a, b)$ meets $M$ in points $(t, a, b)$, which are solutions of $U(t)=0$ (see 1.2). The number of solutions depends on $(a, b)$; sometimes it is 1, sometimes 2 and sometimes 3. The values of $t$ correspond to turning points of the potential energy $V(t)$ (see 1.1), and so to possible stable equilibrium positions for the parabolic lamina.

$M$ divides the $(t, a, b)$-space into two regions: that which is mostly 'above' $M$ has $U(t)>0$ and the other has $U(t)<0$. Minima of $V$ occur for values of $t$ at which $U(t)$ changes from negative to positive. Thus when there is *one* solution to $U(t)=0$ it is a minimum; when there are *three* the middle one is a maximum and the other two are minima; when there are *two*, so that the vertical line touches $M$ ($M$ has vertical tangent plane), the point of contact is neither maximum nor minimum ($U(t)$, hence $V'(t)$ remains the same sign) and the other solution is a minimum. The inner part of the fold gives entirely maxima.

Let us find the set $\mathscr{B}$ in the $(a, b)$-plane consisting of points where there are *two* solutions, i.e. the equation $U(t)=0$ has a repeated root. The condition for $t$ to be a repeated root is $U(t)=U'(t)=0$, i.e.

**1.4**    $2t^3 + t(1-2b) + a = 0$

$6t^2 + (1-2b) = 0$

and elimination of $t$ gives

**1.5**    $27a^2 = 2(2b-1)^3.$

**Fig. 1.3.** The catastrophe surface

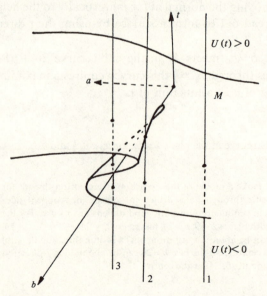

$U(t)>0$

$M$

$a$

$U(t)<0$

$t$

$b$

$3$

$2$

$1$

Note that $a=0$, $b=\frac{1}{2}$ is very special: this gives a triple root, for $U(t)=t^3$ in this case, with triple root 0. The curve $\mathscr{B}$ given by 1.5 is a cuspidal cubic in the $(a, b)$ plane, i.e. in the control space (fig. 1.4).

## 1.6    Exercises

(1) Show that, if $t$ is the repeated root of $U(t)=0$, where $(a, b) \in \mathscr{B}$, then $a=-4t^3$ and $b=\frac{1}{2}(1+6t^2)$. This gives a *parametrization* of $\mathscr{B}$ (with parameter $t$). The points $(t, a, b)$ of $M$ for which $(a, b) \in \mathscr{B}$ therefore have the form $(t, -4t^3, \frac{1}{2}(1+6t^2))$.

This is the curve in $M$ which projects to $\mathscr{B}$ in the $(a, b)$-plane. It is a *space curve* and it has the *regular parametrization* given (this means that the derivatives of the three component functions never vanish for the same $t$ – indeed here the derivative of the first component never vanishes at all). Is the tangent line to the space curve ever vertical?

(2) Sketch the potential function $V$ for various values of $(a, b)$, such as $(0, 0)$, $(0, 1)$, $(1, \frac{1}{2})$. Compare with the rough sketches fig. 1.4. You may find it helpful to use
$$V'(t)=2(1+4t^2)^{-\frac{3}{2}}(2t^3+t(1-2b)-a)$$
and
$$V''(t)=2(1+4t^2)^{-\frac{5}{2}}((16b-2)t^2+12at+1-2b).$$

The set $\mathscr{B}$, which is variously called the *bifurcation set* of $V$ or the *discriminant set* of $U$, separates $(a, b)$ giving one solution to $U(t)=0$ ('outside' $\mathscr{B}$, $>$ in 1.5) from those giving three solutions ('inside' $\mathscr{B}$, $<$ in 1.5). Looking at the surface $M$ from above, a curve like $\mathscr{B}$ is what we 'see' of the folded surface: it is the 'apparent contour' where the surface folds away from us (likewise the apparent contour of a sphere held at arm's length is a circle).

There is another way of regarding this. The first equation of 1.4 is the

Fig. 1.4. Number of solutions of $U(t)=0$, and sketches of $V$ for various positions of $(a, b)$.

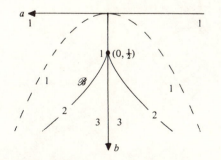
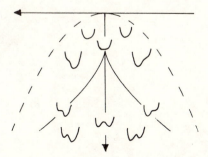

equation of the normal to the parabola $b = a^2$ at $(t, t^2)$: as $t$ varies but $a$ and $b$ remain fixed, it gives the family of all normals to the parabola. Eliminating $t$ between the equations of 1.4 amounts to finding the *envelope* of this family of lines: a curve which is touched by each of the lines of the family (see 1.7 below). The envelope of the normals to a curve is also called the *evolute* of the curve, so $\mathscr{B}$ is the evolute of the parabola $b = a^2$ (fig. 1.5).

### 1.7    Exercise

Show that the curve 1.5 consists precisely of points of the form $(-4t^3, \frac{1}{2}(1 + 6t^2))$ for $t \in \mathbb{R}$. Show that the tangent to 1.5 at the point given by any value of $t$ other than 0 coincides with the normal to the parabola $b = a^2$ at the point $(t, t^2)$. The missing normal, at $(0, 0)$, is the line $a = 0$. Is this in some sense the 'tangent' to 1.5 at $(0, \frac{1}{2})$?

We could go on conjuring more geometry from the parabolic lamina, but let us return to finding stable equilibria. For a given position $(a, b)$ of the centre of mass in the control space we now know (fig. 1.4) how many solutions there are of $U(t) = 0$, i.e. how many stationary points there are of the potential $V$, and also which are minima. Naturally for a physical lamina the centre of gravity will be at some point of the lamina itself, i.e. $(a, b)$ will satisfy $a \leqslant b^2$. The exact values of $t$ can be found by solving the cubic equation $2t^3 + t(1 - 2b) - a = 0$.

It is of particular interest to see what happens to the stable equilibria if we *move* the centre of gravity $(a, b)$ around in control space. We are not concerned here with the dynamics of rolling, merely with the change in the positions of stable equilibrium.

**Fig. 1.5.** Normals to a parabola and their envelope

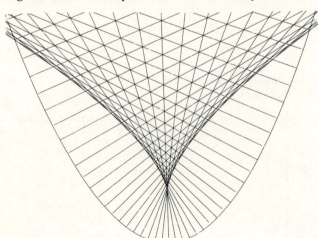

Suppose for example that the point $(a, b)$ moves steadily across $\mathscr{B}$, along the line 12345 in fig. 1.6. Corresponding to 1 there is only *one* stable equilibrium, given by the point 1′ of $M$. At 2 and 3 there are two equilibrium positions available, but in practice the parabola tilts slowly so that positions given by 2′ and 3′ are taken. As $(a, b)$ moves past 4 (which is on $\mathscr{B}$), however, no continuous change in the equilibrium position is possible: the path on $M$ above 345 has a break in it at 4′, where the parabola suddenly spins to the left, a large change in equilibrium position occurring from a small shift in $(a, b)$. This is called a *catastrophe* (fig. 1.7). It is well worth making a model to illustrate this: as described in Poston and Stewart (1978), one can be made from two parallel parabolas made out of thin card, held apart by three or four polystyrene 'pillars'. A magnet and disc of metal on opposite sides of one card can be moved about to change $(a, b)$. Because of the weight of the card, the positions of the *magnet* at which catastrophes occur will not coincide exactly with the bifurcation set $\mathscr{B}$, but can be determined by experiment.

**Fig. 1.6.** Catastrophe surface and its projection to control space

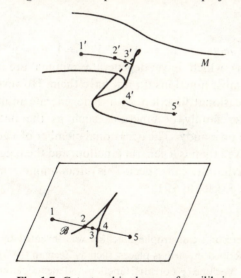

**Fig. 1.7.** Catastrophic change of equilibrium

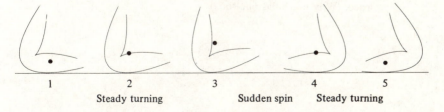

| 1 | 2 | 3 | 4 | 5 |
|---|---|---|---|---|
|  | Steady turning |  | Sudden spin | Steady turning |

## 1.8        Exercises

(1) Describe the behaviour of the lamina as $(a, b)$ moves (*i*) along the path 54321 in fig. 1.6; (*ii*) clockwise round a circle centred at the cusp in fig. 1.6; (*iii*) anticlockwise round the same path. In each case sketch the change in the potential function $V$ as $(a, b)$ traverses the path.

(2) How many normals to the parabola with equation $y = x^2$ pass through a point $(a, b)$ inside the parabola?

(3) A swimmer gets into difficulties in a parabolic cove, and needs to head for the nearest point of land. How many choices does he have? This is similar to, but not identical with, the above example. We should expect him to choose the *absolute* minimum distance always, and the axis of the parabola is now significant. How does the nearest point of shore change as the swimmer's initial position moves across the cove? Does it move continuously, or is there a catastrophe? The reader is recommended *not* to try a practical experiment for this exercise. (This example was suggested to us by Dr I. R. Porteous.)

One final point. Consider the potential function $V$ corresponding to the *cusp point* $(a, b) = (0, \frac{1}{2})$, namely

$$V(t) = \frac{1 + 2t^2}{2(1 + 4t^2)^{\frac{3}{2}}}.$$

This has a *degenerate minimum* at $t = 0$, i.e. $V'(0) = V''(0) = V'''(0) = 0$ but $V^{(4)}(0) \neq 0$. Functions which have degenerate minima are to be regarded as *exceptional*: most functions do not have them. However, we *do* expect to find an occasional function with a degenerate minimum amongst a two-parameter family of functions such as the potential functions $V$ which depend on $a$ and $b$. The occasional member of a general family will be worse behaved than is a general function, and the bigger the family, the worse the behaviour we can expect. This rather ominous maxim will be made a little more precise in 6.21.

## 1.9        Project

An even more interesting catastrophe machine was invented by E. C. Zeeman, and is described in various places, such as Zeeman (1977). Find out about Zeeman's machine, its potential function, and the sudden changes which occur as the control parameters (corresponding to $a$, $b$ above) are varied. (Why not build one too?)

# 2

## *Curves, and functions on them*

'I think, Watson, that you have put on seven and a
half pounds since I last saw you.'
'Seven,' I answered.
'Indeed, I should have thought a little more.
Just a trifle more, I fancy, Watson.'
(*A Scandal in Bohemia*)

Plane curves arise naturally in all sorts of situations and in many guises. Solutions of Newton's laws of motion give the orbits of the planets as ellipses with the Sun at a focus. A spot of paint on a train wheel describes a cycloid as the wheel rolls. These are examples of curves parametrized by time: for each time $t$ a definite point on the curve is determined. If a solid object (such as Dr Watson) is viewed from a distance its outline, also called its apparent contour or profile, is essentially a plane curve (or a curve on the retina), but this time it is not given dynamically as a moving point (fig. 2.1). It is more reminiscent of curves given by equations $f(x, y) = 0$; these latter curves are one of the subjects of chapter 4. A curve may be traced by a linkage of bars and gearwheels; the position of the pencil drawing the curve perhaps depends on the angle of some controlling bar, and so is parametrized by this angle. (Alas! We have no space for this beautiful subject.) When the Sun's rays are reflected from the rounded inner surface of a teacup they produce on the surface of the tea a bright 'caustic' curve. The reflected rays are all tangents to this curve, which is said to be the 'envelope' of the rays. We study envelopes in chapter 5 and caustics at the end of chapter 7.

We are interested also in space curves (curves in $\mathbb{R}^3$) and to a limited

**Fig. 2.1.** Some apparent contours of a torus surface, hidden curves drawn with broken lines

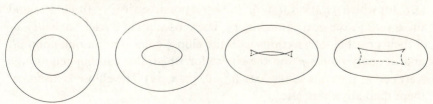

extent in curves in $\mathbb{R}^n$. For the present we look at parametrized curves and coax the geometry from them by means of real-valued functions defined on the curves. To see how this can be geometrically interesting and to set the scene here is an example.

## 2.1    Example

Consider the ellipse $x^2 + 4y^2 = 4$. We can parametrize this by $(2 \cos t, \sin t)$: for each $t$ this point is a definite point on the ellipse, and all points on the ellipse have this form for some $t$. We propose to measure, *at each point* $p_0 = (2 \cos t_0, \sin t_0)$, how 'round' the ellipse is, i.e. how closely it approximates a circle. To do this take a general circle through $p_0$; let its centre be $(a, b)$. Its equation is $C(x, y) = 0$ where

$$C(x, y) = (x - a)^2 + (y - b)^2 - \lambda$$

and the constant $\lambda$ is chosen so that $C$ vanishes at $p_0$. Thus the equation $g(t) = 0$ where

$$g(t) = C(2 \cos t, \sin t) = (2 \cos t - a)^2 + (\sin t - b)^2 - \lambda$$

has one solution $t = t_0$. But $t = t_0$ may be a repeated (double, triple, . . .) solution, which would indicate that several common points of circle and ellipse have come into coincidence at $t_0$. This phenomenon is measured by the *number of derivatives* of $g$ which are zero at $t = t_0$. If $g^{(i)}(t_0) = 0$ for $i = 1, \ldots, k-1$ and $\neq 0$ for $i = k$ then $k$ solutions of $g(t) = 0$ coincide at $t = t_0$; indeed it actually follows that $g(t) = (t - t_0)^k g_1(t)$ for a smooth function $g_1$ with $g_1(t_0) \neq 0$ (compare 3.4). We call this *k-point* contact.

It is easy to verify that $g'(t_0) = 0$ if and only if $-b \cos t_0 + 2a \sin t_0 - 3 \sin t_0 \cos t_0 = 0$, i.e. if and only if $(a, b)$ lies on the normal to the ellipse at $p_0$. For such $(a, b)$ the circle and ellipse touch at $p_0$, having 'two coincident points in common' there. For most $(a, b)$ it is impossible for $g''(t_0)$ to vanish as well: we require $(a, b)$ to satisfy also $b \sin t_0 + 2a \cos t_0 - 3 \cos 2t_0 = 0$. In fact for each $t_0$ there is just one possible $(a, b)$, namely

$$(\tfrac{3}{2} \cos^3 t_0, \ -3 \sin^3 t_0).$$

Thus $(a, b)$ must itself lie on the curve traced out by this point as $t_0$ varies. This curve is called the *evolute* of the ellipse (fig. 2.2). It is the locus of centres of circles having at least 3-point contact with the ellipse and these circles are called *osculating circles* of the ellipse (fig. 2.3). The only points $(a, b)$ for which $g'(t_0) = g''(t_0) = g'''(t_0) = 0$ is possible are $(\pm\tfrac{3}{2}, 0)$, $(0, \pm 3)$, and $\sin 2t_0 = 0$ at the corresponding $t_0$. These osculating circles have *all four* of their possible intersections with the ellipse coincident at one point $p_0$, and the only points $p_0$ on the ellipse for which this can happen are the 'vertices' at the ends of the axes ($t_0 = 0, \tfrac{1}{2}\pi, \pi, \tfrac{3}{2}\pi$). The ellipse is 'rounder' there than anywhere else.

**Fig. 2.2.** Evolute of an ellipse

**Fig. 2.3.** Osculating circles of an ellipse

The function $g$, with the constant $\lambda$ omitted, is called the *distance-squared function on the ellipse*:

$$f(t) = (2 \cos t - a)^2 + (\sin t - b)^2$$

measures the square of the distance from $(a, b)$ to the point with parameter

$t$ on the ellipse. Of course, $f$ and $g$ have the same derivatives. We can put all the distance-squared functions into one family

$$F(t, a, b) = (2 \cos t - a)^2 + (\sin t - b)^2.$$

Such families will be important for us later.

## 2.2     Exercises

(1) In example 2.1 replace the ellipse by the parabola $y = x^2$, parametrised $(t, t^2)$. Find the centres of the osculating circles of the parabola (circles having at least 3-point contact; their centres trace out the evolute of the parabola). Find the unique point $(a, b)$ for which there is a circle with 4-point contact. Note that the equation $f'(t) = 0$ ($f$ defined as in 2.1) has the same solutions as $V'(t) = 0$ ($V$ as in chapter 1).

(2) In example 2.1 replace the ellipse by the ellipse $x^2/A^2 + y^2/B^2 = 1$, parametrised $(A \cos t, B \sin t)$. Show that for each $t$ there is exactly one point $(a, b)$ such that the circle centre $(a, b)$ passing through $(A \cos t, B \sin t)$ has at least 3-point contact there, namely $(a, b) = ((A^2 - B^2) \cos^3 t/A, (B^2 - A^2) \sin^3 t/B)$. Find the points $(a, b)$ for which 4-point contact is possible. What happens when $A = B$?

## Parametrized Curves

In order to talk about general curves we need a proper definition.

## 2.3     Definition

A *(parametrized)* *curve* in the real Euclidean space $\mathbb{R}^n$ is a map

$$\gamma \colon I \to \mathbb{R}^n$$

where $I$ is an open interval in $\mathbb{R}$, where

$$\gamma(t) = (\gamma_1(t), \gamma_2(t), \ldots, \gamma_n(t))$$

and each function $\gamma_i$ has derivatives of all orders, for all $t \in I$. Such a $\gamma_i$ is called *smooth*. The curve $\gamma$ is called *regular* provided there does not exist $t \in I$ with $\gamma'_1(t) = \gamma'_2(t) = \cdots = \gamma'_n(t) = 0$. The variable $t$ is called a *parameter*: the point $\gamma(t)$ has parameter value $t$.

Unless otherwise stated *all curves will be regular* in this book. The vector $(\gamma'_1(t), \gamma'_2(t), \ldots, \gamma'_n(t))$, which is written $\gamma'(t)$ or $d\gamma/dt$, is called the *velocity vector* of $\gamma$ at $t$.

The set of points $\gamma(t)$ in $\mathbb{R}^n$, i.e. the image $C = \gamma(I)$, is what we often think of geometrically as 'the curve'. $C$ is also referred to as the *trace* of $\gamma$. The point $\gamma(t)$ moves along $C$ as $t$ varies, and in fact never stops or turns round, since $\gamma'(t)$ is never zero.

## 2.4      Exercises

Unless otherwise stated, check that the following are regular curves.

(1) $\gamma(t)=(t, t^2)$, $I=\mathbb{R}$ : parabola in $\mathbb{R}^2$.

(2) $\gamma(t)=(\cos t, \sin t)$, $I$ any open interval containing $[0, 2\pi]$: circle in $\mathbb{R}^2$.

(3) $\gamma(t)=(t^3, t^6)$, $I=\mathbb{R}$. The image $\gamma(I)$ is still the parabola of (1), but $\gamma$ is *not* regular at $t=0$. What about $\gamma(t)=(t^2, t^4)$?

(4) $\gamma(t)=(A \cos t, B \sin t)$, $A>0$, $B>0$, $I$ as in (2). The image here is an ellipse in $\mathbb{R}^2$.

(5) $\gamma(t)=(t^2-1, t^3-t)$, $I=\mathbb{R}$. Here the image $\gamma(I)$ is a curve which crosses itself ($t=1, t=-1$ both give $\gamma(t)=(0, 0)$) (fig. 2.4).

(6) $\gamma(t)=(t, t^2, t^3)$, $I=\mathbb{R}$, gives a *twisted cubic curve* in $\mathbb{R}^3$.

(7) $\gamma(t)=(\cos t, \sin t, t)$, $I=\mathbb{R}$, gives a *helix* in $\mathbb{R}^3$.

(8) $\gamma(t)=(t^2, t^3)$, $I=\mathbb{R}$. This is *not* regular at $t=0$, and the image has a *cusp* at $(0, 0)$. (Compare fig. 1.4, where the cusp is at $(0, \frac{1}{2})$.)

(9) The evolute of the ellipse in 2.1, parametrized $\gamma(t)=(\frac{3}{2} \cos^3 t, -3 \sin^3 t)$ is regular except for $t=0, \frac{1}{2}\pi, \pi, \frac{3}{2}\pi$ (and these plus multiples of $2\pi$, which give the same points on the image of $\gamma$). These non-regular points correspond to the cusps on fig. 2.2. What is the corresponding result for the parabola and ellipse of 2.2?

(10) Curves arise naturally as solutions of differential equations. For example, the motion of a particle in a vertical plane under gravity is governed by the equations $y''=-g$, $x''=0$ where $x$ and $y$ are functions of the time $t$ (and we use ' for $d/dt$), and $g$ is a positive constant. These have a solution

$$(x, y)=(at+b, -\tfrac{1}{2}gt^2+ct+d)$$

for any fixed numbers $a, b, c, d$. The formula $\gamma(t)=(x, y)$ gives a curve which is regular for all $t$ unless $a=0$, when it fails to be regular for $t=c/g$. What does this irregularity correspond to physically? For $a\neq 0$ the curve is always a parabola. Note that in an example such as this one the parameter $t$ has a special significance (namely, it is time).

A slightly different kind of solution curve arises for equations such as $y'=ky$ ($k\neq 0$) (where ' means $d/dx$ here) with solution $y=ae^{kx}$ for any fixed

**Fig. 2.4.** The curve of exercise (5)

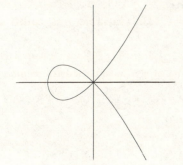

*a.* This curve could conveniently be parametrized by $x$ ($\gamma(x)=(x, ae^{kx})$) but there is nothing wrong with say $\gamma(t)=(t+1, ae^{k(t+1)})$ which has the same image. In both cases we can take $I=\mathbb{R}$.

(11) What is the condition on the constants $a$, $b$, $c$, $d$ for the curve $\gamma(t)$ $=(a\cos t+b\sin t, c\cos t+d\sin t)$ to be regular for all $t$? What is the equation of the curve when it is regular? What happens for other values of $a$, $b$, $c$, $d$?

## Tangent vectors

We shall often use vector notation and draw pictures with vectors. Given points $q$, $r$ in $\mathbb{R}^n$ the segment from $q$ to $r$ represents the vector $r-q$. Thus the segment from 0 to $p$ (0 being the origin) represents the vector $p$ and so does any segment parallel to this (fig. 2.5). Vectors are indicated by arrowed segments in diagrams. The right-hand diagram above indicates the well-known vector law of addition.

WARNING In chapter 4 we shall need to worry about the point where a vector 'starts from' – the vector is said to be *based* at that point. But until then such matters will not concern us.

Let $\gamma: I\rightarrow\mathbb{R}^n$ be a (regular) curve. The vector $v=\gamma(t+h)-\gamma(t)$ corresponds to the chord segment from $\gamma(t)$ to $\gamma(t+h)$.

$$(\gamma_1'(t), \ldots, \gamma_n'(t))=\gamma'(t)=\lim_{h\rightarrow 0}\frac{\gamma(t+h)-\gamma(t)}{h}$$

has for its *direction* the limit of these chords, i.e. the tangent at $\gamma(t)$ (fig. 2.6).

**Fig. 2.5.** Vectors

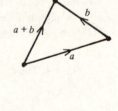

**Fig. 2.6.** Tangent vector

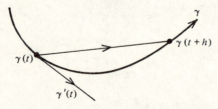

For any $x = (x_1, \ldots, x_n) \in \mathbb{R}^n$ the *length* of $x$ is $\|x\| = (x_1^2 + x_2^2 + \cdots + x_n^2)^{\frac{1}{2}}$. Thus $\|\gamma'(t)\|$ is never zero.

**2.5     Definitions**

The vector $T(t) = \gamma'(t)/(|\gamma'(t)|)$ is called the *unit tangent vector* to $\gamma$ at $t$, or at $\gamma(t)$. The length $\|\gamma'(t)\|$ is called the *speed* of $\gamma$ at $t$. We call $\gamma$ *unit speed* if $\|\gamma'(t)\| = 1$ for all $t$. The *tangent line* to $\gamma$ at $t$ is the straight line through $\gamma(t)$, containing the direction $T(t)$.

Note that $\gamma(t_1) = \gamma(t_2)$ need not imply $\gamma'(t_1) = \gamma'(t_2)$. For example consider $\gamma(t) = (t^2 - 1, t^3 - t)$, $t_1 = -1$, $t_2 = +1$ (fig. 2.4).

For $n = 2$ the equation of the tangent line to $\gamma$ at $t$ is easily verified to be

$$(x_1 - \gamma_1(t))\gamma_2'(t) - (x_2 - \gamma_2(t))\gamma_1'(t) = 0.$$

**2.6     Recall** that the *scalar product* or *dot product* of $x = (x_1, \ldots, x_n)$ and $y = (y_1, \ldots, y_n)$ in $\mathbb{R}^n$ is the real number

$$x \cdot y = x_1 y_1 + x_2 y_2 + \cdots + x_n y_n.$$

We note the following facts:

(1) $x \cdot y = \|x\| \, \|y\| \cos \theta$ where $\theta$ is the angle between the vectors $x$ and $y$ (fig. 2.7). In particular $x \cdot x = \|x\|^2$ and $x \cdot y = 0$ if and only if $x$ and $y$ are *perpendicular*. (The zero vector is perpendicular to everything.)

(2) If the $x_i$ and $y_i$ are all (differentiable) functions of $t$ then

$$(x \cdot y)' = \frac{d}{dt}(x \cdot y) = x \cdot y' + x' \cdot y$$

where $x' = (x_1', \ldots, x_n')$ etc. Also, $\|x\|' = x \cdot x'/\|x\|$.

(3) If $x \cdot x = 1$ (i.e. $x$ is a unit vector) and the $x_i$ are all differentiable functions of $t$, then $x \cdot x' = 0$ for all $t$ so, for each $t$, $x$ and $x'$ are perpendicular vectors.

As a particular case of (3) the unit tangent vector $T$ to a curve $\gamma$ satisfies $T \cdot T' = 0$ so $T'$ is a (possibly zero) vector perpendicular to $T$. (We often write $T$, $T'$ rather than $T(t)$, $T'(t)$; shortness is sometimes better than perfect precision.) Since $\|T\| = 1$, $T'$ measures the rate at which the

**Fig. 2.7.**

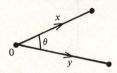

tangent vector is *turning*: the longer $T'$ is the faster $T$ is turning and, roughly speaking, the more curved the curve is.

## Contact

We shall shortly be considering functions defined on curves in $\mathbb{R}^2$ and $\mathbb{R}^3$. Here by way of motivation is a little more on *contact*, which we met previously in 2.1. Let $u$ and $p$ be points of $\mathbb{R}^n$, and consider the two sets of points $x$ of $\mathbb{R}^n$ defined by equations $F(x)=0$ where

**2.7**      $F(x)=\|x-u\|^2-\|u-p\|^2$

**2.8**      $F(x)=(x-p)\cdot u.$

For $n=2$ these are the equations, respectively, of the *circle* centre $u$ passing through $p$ (provided $u\neq p$) and (for $u\neq0$) of the *straight line* through $p$ perpendicular to the vector $u$ (fig. 2.8). For $n=3$ 'circle' becomes *sphere* and 'straight line' becomes *plane*. (For general $n$ the usual names are sphere and hyperplane, respectively.)

In chapter 4 we shall begin to study solution sets of equations, that is sets $F^{-1}(0)=\{x\in\mathbb{R}^n\colon F(x)=0\}$. Granted one condition – that 0 is a 'regular value' of $F$ – we shall find that these sets are very nice, even better than images $\gamma(I)$ of regular curves $\gamma$. For the present we shall not dwell on the technicalities since the above four examples will suffice, but note that the function $F$ used to define the set $F^{-1}(0)$ is important. For example, the square of $F$ will not do (though it defines the same set) because 0 is no longer a 'regular value' of $F^2$. Thus '$F^{-1}(0)$' in what follows assumes that we are using the defining equation $F(x)=0$ (see 2.12(5) for a slight relaxation of this).

Given a (regular) curve $\gamma\colon I\to\mathbb{R}^n$ we measure its contact with $F^{-1}(0)$ as follows:

**Fig. 2.8.**

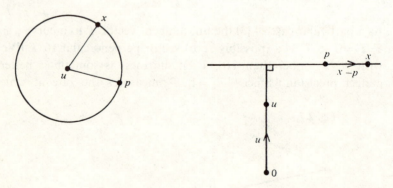

## 2.9 Definition

We say that $\gamma$ and $F^{-1}(0)$ have *k-point* or *k-fold contact* for $t = t_0$ (or, more loosely, at $p = \gamma(t_0)$) provided the function $g$ defined by

$$g(t) = F(\gamma_1(t), \ldots, \gamma_n(t)) = F(\gamma(t))$$

satisfies $g(t_0) = g'(t_0) = \cdots = g^{(k-1)}(t_0) = 0$, $g^{(k)}(t_0) \neq 0$. We also say that the *order of contact* is $k$. Dropping the condition $g^{(k)}(t_0) \neq 0$ we say that there is at least *k*-point contact, or $\geqslant k$-point contact, or that the order of contact is $\geqslant k$. (It might be 'infinite'.)

NOTE We shall always use functions $F$ for which $g$ is a smooth function. Observe that $t_0$, and not just $p$, must be specified.

The condition on $g$ is simply the formulation, in terms of derivatives, that $t_0$ is a $k$-fold root of the equation $g(t) = 0$. It actually implies (see 3.4) that $g(t) = (t - t_0)^k g_1(t)$ for a smooth function $g_1$ with $g_1(t_0) \neq 0$ when $g^{(k)}(t_0) \neq 0$.

As a simple example consider $n = 2$ (plane curves) and lines through $\gamma(t_0)$. The line perpendicular to a unit vector $u$ has equation $(x - \gamma(t_0)) \cdot u = 0$, and the contact with $\gamma$ for $t = t_0$ is measured by $g(t) = (\gamma(t) - \gamma(t_0)) \cdot u$. Now $g'(t_0) = 0$ if and only if $T(t_0) \cdot u = 0$, i.e. $u$ is perpendicular to $T(t_0)$, i.e. the line is parallel to $T(t_0)$ and hence is the tangent line. So our definition does tell us that the tangent line at $t_0$ is the *unique* line having at least 2-point contact for $t = t_0$. (That is certainly good news.) Sometimes tangent lines have much higher contact. For example let $n = 2$, $\gamma(t) = (t, t^k)$, $t_0 = 0$, $F(x_1, x_2) = x_2$, so that $F = 0$ is the $x_1$-axis in $\mathbb{R}^2$. Then $F(\gamma(t)) = t^k$ so the curve has $k$-point contact with the $x_1$-axis at the origin (fig. 2.9). For $k = 1$ the curve (now a line) does not touch the axis at all (1-point contact = crossing without touching) but as $k$ increases, the curve $\gamma$ becomes increasingly 'flat' at the origin. Just as contact with circles (2.1) measures how 'round' the curve is, so contact with lines measures how 'flat' it is.

There are special names for points of plane curves which are especially round or especially flat.

**Fig. 2.9.** $k$-point contact

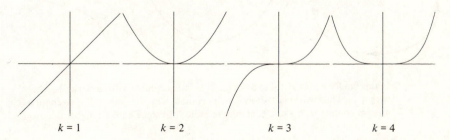

$k = 1$ $\qquad\qquad k = 2 \qquad\qquad k = 3 \qquad\qquad k = 4$

## 2.10      Definition

An *ordinary* (respectively *higher*, or degenerate) *vertex* is a point $p = \gamma(t_0)$ of a plane curve $\gamma$, with parameter value $t_0$, for which there exists a circle* having 4- (resp. at least 5-) point contact with the curve for $t = t_0$. We say $\gamma$ has a vertex at $t_0$, or at $p$.

## 2.11      Definition

An *ordinary* (resp. *higher*, or degenerate) *inflexion* is a point $p = \gamma(t_0)$ of a plane curve $\gamma$, with parameter value $t_0$, for which the tangent line* at $t_0$ has 3- (resp. at least 4-) point contact with the curve for $t = t_0$. We say $\gamma$ has an inflexion at $t_0$, or at $p$.

Thus the ellipse and parabola of 2.1 and 2.2 have no higher vertices, and ordinary vertices at the ends of their axes. (The ellipse in 2.2 (2) with $A = B$ is the exception: if $\gamma$ is a circle then naturally there exists a circle having *very* high contact with $\gamma$ everywhere, namely the circle itself. Every point of a circle is a higher vertex.) See figs. 2.10 and 2.12.

## 2.12      Exercises

(1) Let $\gamma(t) = (t, t^k)$, with $k$ an integer $> 1$. Show that, for any $t_0 \neq 0$ the tangent line to this curve at $t_0$ has exactly 2-point contact with the curve for $t = t_0$. Show that $t_0 = 0$ gives a vertex on the curve if and only if $k = 2$.

(2) Let $\gamma(t) = (t, Y(t))$. Show that the tangent line to the curve $\gamma$ at $t_0$ has $k$-point contact with the curve for $t = t_0$ if and only if $Y^{(i)}(t_0) = 0$ for $2 \leq i \leq k - 1$ and $Y^{(k)}(t_0) \neq 0$. In particular $t = t_0$ gives an inflexion if and only if $Y''(t_0) = 0$, ordinary if and only if also $Y'''(t_0) \neq 0$.

(3) Let $\gamma(t) = (t^2 - 1, t^3 - t)$, $t_0 = 1$, $t_1 = -1$, so that $\gamma(t_0) = \gamma(t_1) = (0, 0)$. Show that the tangent line at $t_0$ has 2-point contact with $\gamma$ for $t = t_0$ and 1-point contact

**Fig. 2.10.** Inflexions and vertices

---

\* In this book we are careful to distinguish between contact with *circles* and contact with *lines*. Thus we do not allow 4-point contact with the tangent line to count as 4-point contact with a circle 'of infinite radius'. In some books a higher inflexion *does* count as a vertex.

with $\gamma$ for $t = t_1$. (The curve has two 'branches' at $(0, 0)$, which cross, and the tangent to one branch does not *touch* the other branch, but crosses it (1-point contact).) Compare fig. 2.4.

(4) (This needs a little real analysis.) Show that the image of a plane curve $\gamma$ crosses its tangent line at an ordinary inflexion.

(5) Let $\gamma : I \to \mathbb{R}^n$ be a (regular) curve. Let $F$ and $F_1$ be functions $\mathbb{R}^n \to \mathbb{R}$ such that the functions $F \circ \gamma$, $F_1 \circ \gamma$ are smooth. Let $t_0 \in I$, $p = \gamma(t_0)$ and suppose $F_1(p) \neq 0$. Show that, for $t = t_0$, the curve $\gamma$ has the same contact with $F^{-1}(0)$ as with $(FF_1)^{-1}(0)$. (You have to verify that $g(t) = F(\gamma(t))$ and $g_1(t) = F(\gamma(t))F_1(\gamma(t))$ have the same number of vanishing derivatives at $t_0$ (remember Leibniz' rule). The result is certainly reasonable, since the set $F(x)F_1(x) = 0$ in $\mathbb{R}^n$ will be, near $p$, *the same* as the set $F(x) = 0$. But note that $F_1 = F$ is definitely *not* allowed.)

(6) Let $\gamma : I \to \mathbb{R}^n$ be a (regular) curve and $L : \mathbb{R}^n \to \mathbb{R}^n$ an invertible linear map. Let $G = F \circ L$, $\delta = L^{-1} \circ \gamma$, where $F : \mathbb{R}^n \to \mathbb{R}$ is such that $F \circ \gamma$ is smooth, as in 2.9. Show that $\delta$ is regular and that $G^{-1}(0)$ and $\delta$ have the same order of contact for $t = t_0$ as $F^{-1}(0)$ and $\gamma$ have for $t = t_0$. (Really this is a triviality. What it says is that *contact is invariant under linear maps such as rotations and reflexions.* The same holds for *translations* $L(x) = x + v$ for a fixed $v$. Illustrate by taking say $n = 2$, $F(x_1, x_2) = x_2$, $\gamma(t) = (t, t^2)$ and various maps $L$ such as $L(x_1, x_2) = (x_1, \lambda x_1 + x_2)$. Draw pictures so that you can see why $G$ has $L$ and $\delta$ has $L^{-1}$.)

Once we have developed a little more in the way of technique we shall return to these geometrical discussions. In practice it is slightly easier to omit certain additive constants from the functions $F(\gamma(t))$ where $F$ is 2.7 or 2.8; of course this makes no difference to derivatives. Here are the details. Let $\gamma : I \to \mathbb{R}^n$ be a (regular) curve.

## 2.13     Definition

Let $u \in \mathbb{R}^n$. The *distance-squared function* on $\gamma$ from $u$ is the function $f_d : I \to \mathbb{R}$ defined by

$$f_d(t) = \| \gamma(t) - u \|^2 = (\gamma(t) - u) \cdot (\gamma(t) - u)$$

Provided $u$ is a unit vector, the *height function* on $\gamma$ in the direction $u$ is the function $f_h : I \to \mathbb{R}$ defined by

$$f_h(t) = \gamma(t) \cdot u$$

Note that $\gamma(t) \cdot u$ is the distance of $\gamma(t)$ from the line (or plane) through 0 perpendicular to $u$ (fig. 2.11), i.e. the 'height of $\gamma(t)$ above this line (or plane)'. We usually take $u$ to be of unit length in the height function: multiplication by a non-zero constant does not affect which derivatives are *zero*.

The following are immediate from the definitions. Let $\gamma : I \to \mathbb{R}^2$ be a (regular) plane curve.

**Fig. 2.11.** Height function

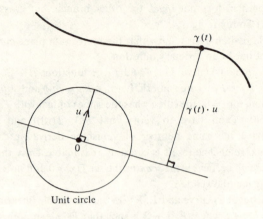

Unit circle

### 2.14    Proposition

$\gamma$ *has k-point contact, for* $t=t_0$, *with the circle, centre u passing through* $\gamma(t_0)$, *if and only if the distance-squared function* $f_d$ *on* $\gamma$ *from u satisfies* $f_d^{(i)}(t_0)=0, i=1,\ldots,k-1; f_d^{(k)}(t_0)\neq0$. *In particular* $\gamma$ *has an ordinary (resp. higher) vertex for* $t=t_0$ *if and only if this holds for* $k=4$ *(resp. some* $k\geqslant5$, *or* $f_d^{(i)}(t_0)=0$ *for all i).*

$\gamma$ *has k-point contact, for* $t=t_0$, *with its tangent line, if and only if the height function* $f_h$ *on* $\gamma$ *in the direction u perpendicular to* $T(t_0)$ *satisfies* $f_h^{(i)}(t_0)=0, i=1,\ldots,k-1; f_h^{(k)}(t_0)\neq0$. *In particular* $\gamma$ *has an ordinary (resp. higher) inflexion for* $t=t_0$ *if and only if this holds for* $k=3$ *(resp. some* $k\geqslant4$, *or* $f_h^{(i)}(t_0)=0$ *for all i).*    □

### 2.15    Exercises

(1) Let $\gamma(t)=(t, Y(t))$ where $Y(0)=Y'(0)=0$. Show that there is an inflexion at $(0,0)$ (parameter value $t=0$) if and only if $Y''(0)=0$, ordinary if and only if also $Y'''(0)\neq0$. Show that there is a vertex at $(0,0)$ (parameter value $t=0$) if and only if $Y''(0)\neq0$ and $Y'''(0)=0$, ordinary if and only if also $Y^{(4)}(0)\neq3(Y''(0))^3$.

(2) With $\gamma$ as in (1) and $Y''(0)\neq0$ find the centre of the (unique) circle which has at least 3-point contact with $\gamma$, for $t=0$. (This is the *osculating circle* or *circle of curvature* there; compare 2.1, 2.27.)

(3) With $\gamma$ as in (1) show that if $Y$ is an *even* function and $Y''(0)\neq0$ then there is a vertex at $(0,0)$ $(t=0)$ and if $Y$ is *odd* then there is an inflexion.

(4) For a fixed real number $\lambda$ let $\gamma(t)=(t^2, \lambda t+t^3)$. Show that $\gamma$ is regular provided $\lambda\neq0$. Show that $t=0$ always gives a vertex of $\gamma$; are there any values of $\lambda$ for which it gives a higher vertex? Investigate the distance-squared functions when $\lambda=0$.

(5) The curve $\gamma(t)=(a \cos t+\cos 2t+1, a \sin t+\sin 2t)$ (polar equation $r=2\cos t+a$, where this is the same $t$ and is the polar angle too), is called a *limaçon* (French for snail, but we cannot see the resemblance). Show that inflexions (resp. vertices) correspond to the parameter values $t$ for which $6a\cos t+a^2+8=0$ (resp. $\sin t(a\cos t+2)=0$). (For the vertices consider the distance-squared function $f_d$ from a general point $(u, v) \in \mathbb{R}^2$ and write the first three derivatives of $f_d$ as linear combinations of $u$, $v$ and 1 (with coefficients which are functions of $t$). Then eliminate by taking the determinant.) For $2<a<4$ there are both vertices and inflexions; for $a=2$ the curve is not regular at $t=\pi$ (it has a cusp there); for $0<a<2$ there is a self-crossing and no inflexions. For $2<a<4$ there are two inflexions which coalesce for $a=4$ into a single higher inflexion – see fig. 2.12. For $a>4$ there are no inflexions. The coalescence and disappearance of inflexions (or indeed vertices) is a common feature of families of curves. At the moment of coalescence something 'worse' happens – a higher inflexion or vertex. (Note that according to our definition the curve for $a=4$ does not have a vertex at $t=\pi$, but has a higher inflexion there. Compare the footnote on p. 18.)

## Reparametrization

There is a slightly unsatisfactory feature of our definition of contact (*now* we admit it!). Consider $\gamma(t)=(t, t^3)$, $t_0=0$, $F(x_1, x_2)=x_2$ so we are looking at contact with the $x_1$-axis, determined by $F(t, t^3)=t^3$: 3-point contact. But $\delta(t)=(1-2t, (1-2t)^3)$ gives 'the same curve' (image $\gamma=$ image $\delta$); it is simply that $\gamma(1-2t)=\delta(t)$ so that for example $\gamma(0)=\delta(\frac{1}{2})$ and $(0, 0)$ has parameter value $\frac{1}{2}$ in $\delta$. Does $\delta$ have the same contact with the $x_1$-axis for $t=\frac{1}{2}$ as $\gamma$ has for $t=0$? Mercifully it does, as you can check easily, but there is clearly a general result we should prove. This changing of parameter is something we can always do without affecting the geometrical curve, and in fact changing parameter substantially shortens some later calculations. So we pause here to give an account of these matters.

The idea is that we make $t$ in $\gamma(t)$ a function of a new parameter $s$, say

**Fig. 2.12.** Inflexions ($\times$) and vertices ($\bullet$) on limaçons

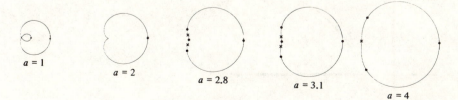

$a = 1$

$a = 2$

$a = 2.8$

$a = 3.1$

$a = 4$

$t = h(s)$. As $s$ traverses some interval $J$ we want $h(s)$ to traverse the interval $I$ on which $\gamma$ is defined, exactly once, without stopping or turning round (fig. 2.13). This just requires $h$ to be monotonic (increasing or decreasing), or equivalently $dt/ds = h'(s)$ to be never zero.

### 2.16     Definition
Let $\gamma: I \to \mathbb{R}^n$ be a (regular) curve. A *change of parameter* for $\gamma$ is a map

$$h: J \to I$$

where $J$ is an open interval, satisfying
(1) $h$ is smooth (i.e. all derivatives $h^{(i)}(s)$ exist for all $s \in J$),
(2) for all $s \in J$, $h'(s) \neq 0$,
(3) $h(J) = I$.
The curve $\delta: J \to \mathbb{R}^n$ given by $\delta(s) = \gamma(h(s))$ is said to be *obtained from $\gamma$ by the change of parameter h.*

### 2.17     Remarks
(1) Is $\delta$ regular? Using the standard calculus rule on each component $\gamma_i \circ h$ of $\delta = \gamma \circ h$ we have $(\gamma_i \circ h)'(s) = \gamma_i'(h(s))h'(s)$; consequently $\delta'(s) = \gamma'(h(s))h'(s)$. This is never zero for $s \in J$, the first factor (a vector) since $\gamma$ is regular, and the second (a number) by (2) above. Of course $\gamma$ and $\delta$ give the same image in $\mathbb{R}^n$. They are different ways of regarding the same set of points in $\mathbb{R}^n$ as a curve.
(2) It follows from 2.15 (1)–(3) that $h$ is a bijection from $J$ to $I$, so that the inverse $h^{-1}: I \to J$ exists. Furthermore $h^{-1}$ is smooth. (This is not quite obvious. First one verifies directly that $(h^{-1})'(t) = 1/h'(h^{-1}(t))$ for all $t \in I$, using the (crucial) fact that $h'(s)$ is always non-zero. Then, knowing $h^{-1}$ to have a derivative, we can

**Fig. 2.13.** Change of parameter

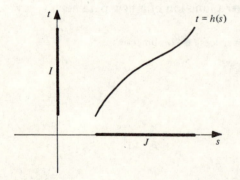

write down $(h^{-1})''(t)$ by standard calculus rules, and so on by induction.) In fact the requirements (1)–(3) of 2.15 are equivalent to (1), (3) and the following condition: $h^{-1}$ (which exists by (1) and (3)) is smooth.

(3) A map $h$ as in 2.15 is called a *diffeomorphism* from $J$ to $I$. By the remarks above, $h^{-1}$ is then a diffeomorphism from $I$ to $J$. Also the composite of two diffeomorphisms is a diffeomorphism.

(4) Suppose that $f$ and $g$ are smooth functions $I \to \mathbb{R}$ ($I$ an open interval, as usual). Then $f \pm g$, $fg$ and $cf$ ($c$ constant) are all smooth. Furthermore if $g(t)$ is never zero then $f/g$ and $g^{1/n}$ ($n$ odd) are smooth. For $n$ even one needs $g(t)$ to be $>0$ rather than just nonzero. All of these are easy to prove from standard calculus rules.

## 2.18    Exercises

(1) Show that $h(s) = s^2$ defines a diffeomorphism from $(1, 2)$ to $(1, 4)$, but not from $(-1, 2)$ to $(1, 4)$.

(2) $h(s) = s^3$ does not define a diffeomorphism from $(-1, 1)$ to $(-1, 1)$, even though $h^{-1}$ exists here.

(3) $h(s) = as + b$ ($a$, $b$ constants $a \neq 0$) defines a diffeomorphism from $I$ to $h(I)$ for any open interval $I$.

(4) $h(s) = e^s$ defines a diffeomorphism from $I$ to $h(I)$ for any open interval $I$.

(5) Let $a$, $b$, $c$, $d$, $e$ be constants with $d < e$ and let $h(s) = as^2 + bs + c$, $I = (d, e)$. Find the conditions on $a$, $b$, $c$, $d$, $e$ which ensure that $h$ is a diffeomorphism from $I$ to $h(I)$.

(6) Let $\gamma : I \to \mathbb{R}^2$ be a curve and $ax + by + c = 0$ a fixed line in $\mathbb{R}^2$. Define $f : I \to \mathbb{R}$ by $f(t) = $ the square of the distance from $\gamma(t)$ to the line. Verify that $f$ is smooth. Further, show that if $a$, $b$ and $c$ are smooth functions of $t$ with $a(t)$ and $b(t)$ never both zero for the same $t$, then the square of the distance from $\gamma(t)$ to this variable line also gives a smooth function. What about the distance function? The signed distance function? (Use the rules in 2.17 (4) above.)

Recall from 2.9 that contact between a curve $\gamma$ and a 'smooth hypersurface' $F(x) = 0$ in $\mathbb{R}^n$ is measured by the number of vanishing derivatives of $g$, where $g(t) = F(\gamma(t))$. (In some special cases 2.7, 2.8, it is also measured by the number of vanishing derivatives of distance-squared or height functions.) Under a change of parameter $h$, $g$ becomes $g \circ h$. Thus, the following result shows in particular that $\gamma$ *and any reparametrization* $\delta = \gamma \circ h$ *of* $\gamma$ *have the same contact with* $F(x) = 0$.

## 2.19    Proposition

*Suppose that* $g : I \to \mathbb{R}$ *is smooth and* $h : J \to I$ *is a diffeomorphism. Let* $t_0 \in I$, $s_0 = h^{-1}(t_0)$. *Then the derivatives*

$$g^{(i)}(t_0), \quad i = 1, 2, \ldots, k, \text{ are all zero}$$

*if and only if the derivatives*

$$(g \circ h)^{(i)}(s_0), \ i = 1, 2, \ldots, k, \ are \ all \ zero.$$

**Proof** We have $(g \circ h)'(s) = g'(h(s))h'(s)$.
By repeatedly differentiating this formula we obtain

$$(g \circ h)^{(p)}(s) = g^{(p)}(h(s))(h'(s))^p + \text{terms having as factors}$$
$$g^{(i)}(h(s)) \text{ for } i = 1, \ldots, p-1.$$

It follows that if $g^{(i)}(h(s_0)) = 0$ for $i = 1, \ldots, p$, then $(g \circ h)^{(p)}(s_0) = 0$. Applying this for $p = 1, 2, \ldots, k$ gives the result in one direction; the other direction is obtained by writing $g = (g \circ h) \circ h^{-1}$ and applying the same method.    □

### 2.20    Remark

It also follows that (with $g$, $h$ as in 2.19) $g$ has a local maximum (resp. minimum) at $t$ if and only if $g \circ h$ has a local maximum (resp. minimum) at $s_0$. (Proof: $g$ has a local maximum at $t_0$ if and only if $g'(t_0) = 0$ and $g'(t)$ changes from positive to negative as $t$ increases through $t_0$. The result then follows from the first line of the proof of 2.19 – note that if $h'(s_0) < 0$ then $g'(h(s))$ changes from negative to positive as $s$ *increases* through $s_0$, but the factor $h'(s)$ reverses the sign.)

For a plane curve $\gamma: I \rightarrow \mathbb{R}^2$, and a given $t_0 \in I$, let $l(t)$ be the arclength of the curve $\gamma(I)$ from $\gamma(t_0)$ to $\gamma(t)$. With $\gamma(t) = (X(t), Y(t))$ we have

$$\left(\frac{dl}{dt}\right)^2 = \left(\frac{dX}{dt}\right)^2 + \left(\frac{dY}{dt}\right)^2$$

so that

$$l(t) = \int_{t_0}^{t} \left[\left(\frac{dX}{dt}\right)^2 + \left(\frac{dY}{dt}\right)^2\right]^{\frac{1}{2}} dt = \int_{t_0}^{t} \|\gamma'(t)\| \, dt$$

where we agree that $l(t)$ should be $> 0$ for $t > t_0$ and $< 0$ for $t < t_0$. The intuitive idea is suggested in fig. 2.14; for more discussion see for example Willmore (1959). It is the same for curves in $\mathbb{R}^n$.

**Fig. 2.14.** Arclength

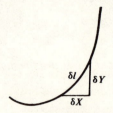

## 2.21    Definition

The *arc-length* of a (regular) curve $\gamma: I \to \mathbb{R}^n$, measured from $\gamma(t_0)$, where $t_0 \in I$, is

$$l(t) = \int_{t_0}^{t} \|\gamma'(u)\| \, du \qquad (t \in I)$$

In particular if $\gamma$ is unit speed (i.e. $\|\gamma'(t)\| = 1$ for all $t$), then $l(t) = t - t_0$, i.e. apart from an additive constant the parameter $t$ measures arc-length. A unit speed curve is often said to be *parametrized by arc-length*.

Arc-lengths are more or less impossible to work out explicitly, except in very special examples. In case the reader insists on working out an example, here are a few possibilities.

| | | |
|---|---|---|
| $n=2$ | $\gamma(t) = (t - \sin t, 1 - \cos t)$ | (cycloid) |
| | $\gamma(t) = (t, \log t)$ | (graph of log function, $t > 0$) |
| | $\gamma(t) = (t, t^2)$ | (parabola: this is hard) |
| | $\gamma(t) = (R \cos t, R \sin t)$ | (circle radius $R$: this is easy) |
| $n=3$ | $\gamma(t) = (\cos t, \sin t, t)$ | (helix) |

The importance of arc-length for us is the following. If we reparametrize a given curve $\gamma$ using arc-length from a fixed $t$ as the new parameter, then the curve will become unit speed.

Let $\gamma: I \to \mathbb{R}^n$ be a (regular) curve, and let $t_0 \in I$. Define $l(t)$ as in 2.21. Then $l$ is smooth and $l'(t) = \|\gamma'(t)\| > 0$ for all $t \in I$, so $l$ defines a diffeomorphism from $I$ to an open interval $J$, namely $J = (l(a), l(b))$ where $I = (a, b)$. Let $l^{-1}$ be the inverse of $l$; then $l^{-1}$ is also a diffeomorphism by 2.17 (2), i.e. $l^{-1}$ is a change of parameter for the curve $\gamma$.

## 2.22    Proposition

*The curve* $\alpha = \gamma \circ l^{-1}: J \to \mathbb{R}^n$ *is unit speed.*

**Proof** Write $h = l^{-1}$. Then $(\gamma \circ h)'(s) = \gamma'(h(s))h'(s)$ for all $s \in J$.
However differentiating $l(h(s)) = s$ gives $l'(h(s))h'(s) = 1$ and $l'(h(s)) = \|\gamma'(h(s))\|$
By definition of $l$. Hence $\|(\gamma \circ h)'(s)\| = 1$, as required. $\qquad\qquad\square$

Again only rather feeble examples can be written down explicitly. The circle, radius $R$, centre $(0, 0)$ has parametrization $\gamma(t) = (R \cos t, R \sin t)$ and unit speed parametrization $\alpha(s) = \gamma(s/R) = (R \cos (s/R), R \sin (s/R))$. The other examples above can be treated similarly.

### Curvature

Now that we know any regular curve can be made unit speed by a mere reparametrization, we shall often assume our curves to be unit speed. When it makes a difference to the formulae we shall say so. To make

LIBRARY
MOUNT ST. MARY'S
COLLEGE
EMMITSBURG, MARYLAND

matters (we hope) clearer, unit speed curves will be denoted by $\alpha$ below, up to 2.29, and their parameter will be called $s$. Thus $\alpha'(s) = T(s)$ for any unit speed $\alpha$ (compare 2.5); for any $\gamma$, $T(t) = \gamma'(t)/\|\gamma'(t)\|$.

### 2.23     Curvature and normal ($n \geqslant 3$)

Let $\alpha: I \to \mathbb{R}^n$ be unit speed, and let $n \geqslant 3$. *The curvature of $\alpha$ at $s$ is defined to be* $\kappa(s) = \|T'(s)\| = ((\alpha_1''(s))^2 + \cdots + (\alpha_n''(s))^2)^{\frac{1}{2}}$. Thus $\kappa(s) \geqslant 0$ and $\kappa$ is a smooth function when restricted to the (open) subset of $I$ where $\kappa(s) \neq 0$. If $\kappa(s) \neq 0$ then the *principal normal $N(s)$ is the unit vector* $T'(s)/\kappa(s)$. Since $T(s)$ is a unit vector, $N(s)$ is perpendicular to $T(s)$. When $\kappa(s) = 0$, $N(s)$ is *not defined*.

For any arbitrary (regular) $\gamma$, with arclength function $l$ (relative to some $t_0 \in I$), we define $\kappa(t)$ to be the curvature of the unit speed curve $\alpha = \gamma \circ l^{-1}$ at $s = l(t)$ (see 2.22). This does not depend on the choice of $t_0$, and we have $\kappa(t) = \|T'(t)\|/l'(t) = \|T'(t)\|/\|\gamma'(t)\|$.

(***Proof*** Write $T(t)$ for the unit tangent to $\gamma$ at $t$ and $T_\alpha(s)$ for the unit tangent to $\alpha$ at $s$; then $T(t) = T_\alpha(l(t))$. Thus $T'(t) = T_\alpha'(l(t))l'(t)$ and taking lengths gives $\|T'(t)\| = \kappa_\alpha(l(t))l'(t) = \kappa(t)l'(t)$. Perhaps it is 'clearer' in old-fashioned notation: write $s = l(t)$; then

$$\frac{dT}{ds}\frac{ds}{dt} = \frac{dT}{dt}$$

and the length of the first term on the left is $\kappa$.)

### 2.24     Curvature and normal (plane curves)

When $n = 2$ we can usefully give the curvature a sign. Let $\alpha: I \to \mathbb{R}^2$ be unit speed; then $T$ is as before, but this time define the *unit normal $N(s)$* to be obtained from $T(s)$ by rotating anticlockwise through $\frac{1}{2}\pi$. If $\alpha(s) = (X(s), Y(s))$ then $T(s) = (X'(s), Y'(s))$ and $N(s) = (-Y'(s), X'(s))$. Now $T'(s)$ is still perpendicular to $T(s)$ so (as we are in the plane) there is a real number $\kappa(s)$ such that

$$T'(s) = \kappa(s)N(s).$$

We call $\kappa(s)$ the *curvature* of $\alpha$ at $s$. Note $\kappa(s) = \pm\|T'(s)\|$. With $\alpha$ as above we have, omitting $s$,

$$T' = \alpha'' = (X'', Y'') = \kappa N.$$

Dotting both sides with $N = (-Y', X')$ we have

$$\kappa = X'Y'' - X''Y'.$$

In particular $\kappa$ is a smooth function of $s$ for all $s \in I$. Remember this formula assumes unit speed (see below).

Note that $\kappa(s) > 0$ if $T(s)$ is turning towards the normal $N(s)$ and $\kappa(s) < 0$ if $T(s)$ is turning away from $N(s)$. See fig. 2.15.

Suppose that the angle between the tangent and the x-axis is $\psi(s)$ (see the right-hand figure), so that $\cos\psi = X'$, $\sin\psi = Y'$. (It is not too hard to show that there is a *smooth* function $\psi: I \to \mathbb{R}$ with these properties, which is unique once we fix the angle for a particular $s_0 \in I$ to be, say, in the range $0 \leqslant \psi(s_0) < 2\pi$). Then $X'' = (-\sin\psi)\psi'$, $Y'' = (\cos\psi)\psi'$ so $\kappa(s) = \psi'(s)$. This again assumes unit speed.

For an arbitrary (regular) $\gamma: I \to \mathbb{R}^2$, $\gamma(t) = (X(t), Y(t))$ define $\kappa(t)$ as the curvature of the unit speed curve $\gamma \circ l^{-1}$ at $l(t)$. We have

$$T(t) = (X'(t), Y'(t))/((X'(t))^2 + (Y'(t))^2)^{\frac{1}{2}}$$
$$N(t) = (-Y'(t), X'(t))/((X'(t))^2 + (Y'(t))^2)^{\frac{1}{2}},$$

the denominators being of course $l'(t)$ ($l = $arclength from some $t_0$). As for the curvature, we have $\gamma = \alpha \circ l$ as in 2.22 with $\alpha$ unit speed. Thus

$$\gamma'(t) = \alpha'(l(t))l'(t) = T(t)l'(t)$$
$$\gamma''(t) = \alpha''(l(t))(l'(t))^2 + \alpha'(l(t))l''(t).$$

Now $\alpha$ is unit speed, so $\alpha''(l(t))$ is the curvature times the unit normal, and $\alpha'(l(t))$ is the unit tangent. Dotting both sides of the last equation with $N(t)$ ($= N_\alpha(l(t))$ where $N_\alpha$ is the normal for $\alpha$) we have

$$\gamma''(t) \cdot N(t) = \kappa(t)(l'(t))^2$$

and substitution gives, dropping $t$ now,

$$\kappa = \frac{X'Y'' - X''Y'}{(X'^2 + Y'^2)^{\frac{3}{2}}}$$

The tangent angle formula becomes $\kappa(t) = \psi'(t)/l'(t)$.

## 2.25    Serret–Frenet formulae for plane curves

These are formulae for $T'$, $N'$ in terms of $T$, $N$. For $\alpha$ unit speed we already have, omitting $s$,

$$T' = \kappa N$$

(i.e. $T'(s) = \kappa(s) N(s)$ for all $s \in I$) from 2.24. Now $N'$ will be perpendicular

**Fig. 2.15.**

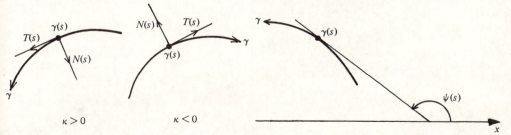

$\kappa > 0$          $\kappa < 0$

to $N$ (see 2.6 (3)) and so $N' = \lambda T$ for some real $\lambda$ (depending on $s$). Also $T \cdot N = 0$, so $T'N + TN' = 0$, which gives $\kappa + \lambda = 0$. Hence

$$N' = -\kappa T.$$

For arbitrary plane curves we obtain

$$T' = \kappa N l' = \kappa N \|\gamma'\|$$
$$N' = -\kappa T l' = -\kappa T \|\gamma'\|.$$

### 2.26    Exercises

(1) For (1), (2), (4), (5), (6), (7) and the regular points of (8), (9) in (2.4) find formulae for the unit tangent, unit normal (or principal normal) and curvature as functions of $t$. (For example, for the ellipse in (4) we get

$$T(t) = (-A \sin t, B \cos t)/(A^2 \sin^2 t + B^2 \cos^2 t)^{\frac{1}{2}}$$
$$N(t) = (-B \cos t, -A \sin t)/(A^2 \sin^2 t + B^2 \cos^2 t)^{\frac{1}{2}}$$
$$\kappa(t) = AB/(A^2 \sin^2 t + B^2 \cos^2 t)^{\frac{3}{2}}.)$$

(2) Let $\gamma(t) = (t, Y(t))$. Show that $\kappa(t) = Y''(t)/(1 + (Y'(t))^2)^{\frac{3}{2}}$.

(3) The cubic curve $y^2 = x - x^3$ consists of an oval, lying in the region $0 \leqslant x \leqslant 1$, and another part (fig. 2.16). Removing the points $(0, 0)$ and $(1, 0)$ the oval splits into two pieces, parametrized by $\gamma(t) = (t, \pm(t - t^3)^{\frac{1}{2}})$. Show that, for these curves $(0 < t < 1)$, $\kappa(t) = \pm 2(3t^4 - 6t^2 - 1)/(9t^4 - 4t^3 - 6t^2 + 4t + 1)^{\frac{3}{2}}$. Why do the two halves have opposite signs for $\kappa$? Is $\kappa$ positive or negative for the top half? (And is that the upper or the lower sign?)

**Fig. 2.16.** The curve of example (3)

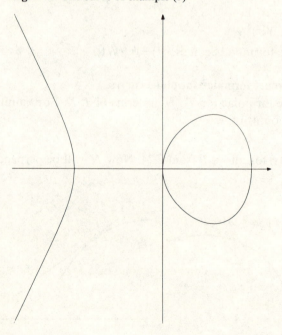

(4) Let $\gamma: I \to \mathbb{R}^n$ be a (regular) curve and $\delta = \gamma \circ h$ where $h: J \to I$ is a change of parameter. Writing $\kappa_\gamma$, $\kappa_\delta$ for the curvatures, show $\kappa_\gamma(t) = \pm \kappa_\delta(h^{-1}(t))$ for all $t \in I$, $+$ if $h'(s)$ is always $>0$, $-$ if $h'(s)$ is always $<0$. (This just says that curvature is independent up to sign of parametrization, and is a tiny extension of the definition. In fact $\gamma$ and $\delta$ can both be reparametrized to the same unit speed curve $\alpha: K \to \mathbb{R}^n$ or one to $\alpha$ and one to $-\alpha$, where $-\alpha$ has domain $-K = \{-u : u \in K\}$ and formula $-\alpha(-u) = \alpha(u)$.)

(5) Let $\gamma(t) = (t^r, t^s)$ where $t > 0$ and $s > r > 0$. Show that $\gamma$ is regular and find a formula for $\kappa(t)$. Show that, as $t \to 0$, $\kappa(t) \to 0$ when $s > 2r$ and $\kappa(t) \to \infty$ when $s < 2r$. (The image of $\gamma$ for all (small) $t$ is called an *ordinary cusp* when $r = 2$, $s = 3$ (here $\kappa(t) \to \infty$) and a *rhamphoid cusp* when $r = 2$, $s = 5$ (here $\kappa(t) \to 0$). More about cusps in chapter 6 – see 6.19.)

## Functions on plane curves

We can now use the formulae of 2.24 to discuss in detail the distance-squared and height functions introduced in 2.13 and hence contact of plane curves with circles and lines. Recall from 2.19 that reparametrization makes no difference to the number of vanishing derivatives of a function on a curve, so there really is no loss of generality in considering only unit speed curves here.

### 2.27 Distance-squared functions on plane curves

Let $\alpha: I \to \mathbb{R}^2$ be unit speed, and $u = (a, b) \in \mathbb{R}^2$. We consider the distance-squared function

$$f(s) = \|\alpha(s) - u\|^2 = (\alpha(s) - u) \cdot (\alpha(s) - u).$$

Writing $\alpha$ for $\alpha(s)$, etc., we have

$$\tfrac{1}{2} f'(s) = (\alpha - u) \cdot \alpha' = (\alpha - u) \cdot T$$
$$\tfrac{1}{2} f''(s) = \alpha' \cdot T + (\alpha - u) \cdot T' = 1 + (\alpha - u) \cdot \kappa N$$
$$\tfrac{1}{2} f'''(s) = \alpha' \cdot \kappa N + (\alpha - u) \cdot \kappa' N + (\alpha - u) \cdot \kappa N'$$
$$= (\alpha - u) \cdot \kappa' N - (\alpha - u) \cdot \kappa^2 T,$$

using $T \cdot T = 1$, $T \cdot N = 0$. It is now a simple matter to write down conditions for the first few derivatives to vanish.

We find

$f'(s) = 0 \Leftrightarrow \alpha - u = \lambda N (\lambda \in \mathbb{R}) \Leftrightarrow u$ is on the normal line to $\alpha$ at $s$
$f'(s) = f''(s) = 0 \Leftrightarrow \kappa \neq 0$ and $\alpha - u = -N/\kappa \Leftrightarrow \kappa \neq 0$ and $u = \alpha + N/\kappa$ at $s$
$f' = f'' = f''' = 0$ at $s \Leftrightarrow \kappa \neq 0$, $u = \alpha + N/\kappa$ and $\kappa' = 0$ at $s$.

Working out $f^{(4)}$ we find that the extra condition for $f^{(4)}$ to vanish at $s$ as well is that $\kappa''(s) = 0$.

Various deductions can be made from this and 2.14.

(i) Provided $\kappa(s) \neq 0$, there is a unique circle having at least 3-point contact with $\alpha$ at $s$. The circle is called the *osculating circle* or *circle of curvature* at $s$; its centre $\alpha(s) + N(s)/\kappa(s)$ is called the *centre of curvature* of $\alpha$ at $s$ (or more loosely at $\alpha(s)$); its radius $\rho(s) = 1/|\kappa(s)|$ the *radius of curvature*. See fig. 2.3 for some circles of curvature of an ellipse. (The case $\kappa(s) = 0$ appears in 2.29.)

(*ii*) The unique circle of (i) has at least 4-point contact (resp. at least 5-point contact) iff $\kappa'(s) = 0$ (resp. $\kappa'(s) = \kappa''(s) = 0$); hence this is the necessary and sufficient condition (granted $\kappa(s) \neq 0$) for a vertex* (resp. higher vertex) of $\alpha$.

(*iii*) An ordinary vertex occurs precisely at a simple maximum or minimum of curvature, where $\kappa(s) \neq 0$.

For arbitrary (regular) plane curves $\gamma$ all of the above results hold. (Reparametrize $\gamma$ as $\alpha = \gamma \circ l^{-1}$ (2.22); this does not affect contact. As for the conditions $\kappa'(s) = 0$, etc., recall $\kappa_\gamma(t) = \kappa_\alpha(l(t))$ by definition 2.24, so that $\kappa'_\gamma(t) = \kappa'_\alpha(l(t))l'(t)$ and $\kappa'_\gamma(t) = 0$ iff $\kappa'_\alpha(l(t)) = 0$, etc.)

## 2.28    Exercises

(1) Let $\alpha$ be a unit speed plane curve and let $f$ be the distance-squared function as in 2.27. Suppose $f'(s) = 0$, so that $u = \alpha(s) + \lambda N(s)$ for some $\lambda \in \mathbb{R}$, and also suppose $\kappa(s) > 0$. Show, by investigating the sign of $f''(s)$, that $f$ has a *maximum* at $s$ if $\lambda > 1/\kappa(s)$ and a *minimum* if $\lambda < 1/\kappa(s)$. Thus the centre of curvature at $s$ separates points on the normal for which the distance to the curve (or its square) is minimized at $\alpha(s)$ from those for which it is maximized. Can you say anything about $\lambda = 1/\kappa(s)$?

(2) Let $\alpha$ be a unit speed plane curve and define

$$g(s) = (\alpha(s) - u) \cdot N(s),$$

for a fixed $u \in \mathbb{R}^2$. Show that $g(s)$ is, up to sign, the distance from $u$ to the tangent to $\alpha$ at $s$. Show the following:

(i) $2g'(s) = -\kappa(s)f'(s)$ where $f$ is as in 2.27,

(ii) $g'(s) = 0 \Leftrightarrow \kappa(s) = 0$ or $u$ lies on the normal at $s$,

(iii) $g'(s) = g''(s) = 0 \Leftrightarrow (\kappa(s) = \kappa'(s) = 0)$ or $(\kappa(s) = 0$ and $u$ lies on the normal at $s$) or ($u$ is the centre of curvature at $s$).

What is the condition for $g'(s) = g''(s) = g'''(s) = 0$? What does it become if $\kappa(s) \neq 0$? (Compare the Introductory Example in chapter 1.)

(3) Let $\alpha$ be a unit speed plane curve and define

$$g(s) = (\alpha(s) - u) \cdot T(s).$$

What distance does $g(s)$ represent? Find the conditions for the first $k$ derivatives of $g$ to vanish at $s$, for $k = 1, 2, 3$.

---

* In some books a point where $\kappa = \kappa' = 0$ counts as a vertex. See the footnote on p. 18.

(4) Let $\alpha$ be a unit speed plane curve, with $\alpha(0)=(0, 0)$, $\alpha'(0)=T(0)=(1, 0)$, as in fig. 2.17. Show that

$$\alpha''(0)=(0, \kappa(0))$$
$$\alpha'''(0)=(-\kappa^2(0), \kappa'(0))$$
$$\alpha^{(4)}(0)=(-3\kappa(0)\kappa'(0), \kappa''(0)-(\kappa(0))^3).$$

(5) Let $\alpha: I \to \mathbb{R}^2$ be unit speed and let $F: \mathbb{R}^2 \to \mathbb{R}^2$ be a linear map with matrix $A$ (relative to standard bases). Let $\beta = F \circ \alpha$. Show that $\beta'(s)=A\alpha'(s)$ (where we write $\alpha'$, $\beta'$ as column vectors) and deduce that $\beta$ is regular provided $A$ is nonsingular, unit speed provided $A$ is orthogonal. Deduce that, if $A$ is orthogonal, then the curvatures of $\alpha$ and $\beta$ are related by $\kappa_\beta(s)= \pm \kappa_\alpha(s)$, the sign being that of $\det(A)$. This shows that *rotations preserve curvature, reflections reverse it in sign.* (Obviously, translations preserve curvature: here, $\beta(s)=\alpha(s)+v$ for a fixed vector $v$.).

(6) Let $\gamma$ be a plane curve with arclength function $l$, and $\alpha = \gamma \circ l^{-1}$ be unit speed as in 2.22. Show that $\kappa'(t_0)= \cdots = \kappa^{(k)}(t_0)=0 \Leftrightarrow \kappa_\alpha'(s_0)= \cdots = \kappa_\alpha^{(k)}(s_0)=0$, where $s_0 = l(t_0)$. Remember $\kappa(t)=\kappa_\alpha(l(t))$ by definition.

(7) Consider curves parametrized $\gamma(t)=(t, Y(t))$ for some $Y$ with $Y(0)=Y'(0)=0$. Let $\alpha = \gamma \circ l^{-1}$ (where $l(0)=0$) be the unit speed curve corresponding to $\gamma$ and $\kappa_\alpha(s)=\kappa(l^{-1}(s))$) be the curvature of $\alpha$ (see 2.24). Show that, for any $p \geqslant 2$, any one of the sets of numbers

   (i) $Y''(0), Y'''(0), \ldots, Y^{(p)}(0)$

   (ii) $\kappa(0), \kappa'(0), \ldots, \kappa^{(p-2)}(0)$

   (iii) $\kappa_\alpha(0), \kappa_\alpha'(0), \ldots, \kappa_\alpha^{(p-2)}(0)$

uniquely determines the the other two. (Repeated differentiation of $Y''=\kappa(1+Y'^2)^{\frac{3}{2}}$ shows that $Y^{(p)}=\kappa^{(p-2)}(1+Y'^2)^{\frac{3}{2}}+$ (terms involving lower derivatives of $Y$ and $\kappa$), and this shows that (i) and (ii) determine each other. Repeated differentiation of $l'^2=1+Y'^2$ shows that $Y''(0), \ldots, Y^{(p-1)}(0)$ determine $l'''(0), \ldots, l^{(p)}(0)$ (note $l'(0)=1, l''(0)=0$ automatically). Using this and $\kappa_\alpha(l(t))=\kappa(t)$ shows that (i) and (ii), which determine each other, also determine (iii). Finally (iii) gives (ii) once $l'''(0), \ldots, l^{(p-2)}(0)$ are known, and these come inductively, determining them from $Y''(0), \ldots, Y^{(p-3)}(0)$ and these from $\kappa(0), \ldots, \kappa^{(p-5)}(0)$.)

**Fig. 2.17.**

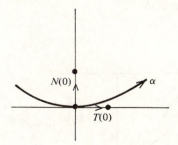

Deduce that $\gamma$ has $\geqslant k$-point contact with the $x_1$-axis in $\mathbb{R}^2$ $(k \geqslant 3)$ iff $\kappa(0) = \cdots \kappa^{(k-3)}(0) = 0$. Deduce also that *any* regular curve $\gamma: I \to \mathbb{R}^2$ has $\geqslant k$-point contact with its tangent line at $\gamma(t_0)$ iff $\kappa(t_0) = \cdots = \kappa^{(k-3)}(t_0) = 0$.

(8) In the notation of (7), show that the first few formulae work out as follows:
$$\kappa(0) = \kappa_\alpha(0) = Y''(0), \; \kappa'(0) = \kappa'_\alpha(0) = Y'''(0), \; \kappa''(0) = \kappa''_\alpha(0) = Y^{(4)}(0) - 3(Y''(0))^3.$$
(Compare 2.15(1).)

## 2.29     Height functions on plane curves   (see 2.13)

This proceeds much as for distance-squared functions. Let $\alpha: I \to \mathbb{R}^2$ be unit speed, and consider, for a fixed unit vector $u \in \mathbb{R}^2$,
$$f(s) = \alpha(s) \cdot u$$
Thus $f'(s) = T(s) \cdot u$, $f''(s) = -\kappa(s) N(s) \cdot u$, etc. We find

$f'(s) = 0$ $\qquad\qquad \Leftrightarrow N(s) = \pm u$ (since $N$, $u$ are unit vectors)

$f'(s) = f''(s) = 0$ $\qquad \Leftrightarrow N(s) = \pm u$ and $\kappa(s) = 0$

$f'(s) = f''(s) = f'''(s) = 0 \Leftrightarrow N(s) = \pm u$ and $\kappa(s) = \kappa'(s) = 0$

If you go on long enough, you can re-prove (7) of the exercises above. Note, at any rate, that *the necessary and sufficient condition for $\alpha$ to have an ordinary (resp. higher) inflexion for $s = s_0$ is that $\kappa(s_0) = 0$, and $\kappa'(s_0) \neq 0$* (resp. $\kappa'(s_0) = 0$). This result holds whether $\alpha$ is unit speed or not.

Now that we are not juxtaposing results for unit speed and other curves we shall often use $\gamma$, parameter $t$, for a unit speed curve, but always say when it *is* assumed unit speed.

## 2.30     Exercises

(1) **Pedal** Show that the foot of the perpendicular from the origin to the tangent to the plane curve $\gamma$ at $\gamma(t)$ is given by $(\gamma(t) \cdot N(t)) N(t)$ (fig. 2.18). As $t$ varies, this point moves on a curve called the *pedal curve* of $\gamma$ with respect to 0. Let us write
$$\delta(t) = (\gamma(t) \cdot N(t)) N(t), \text{ or just } \delta = (\gamma \cdot N) N$$

**Fig. 2.18.** Pedal curve

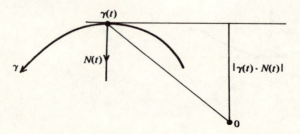

and now assume $\gamma$ is unit speed ($\delta$ won't be, in general). Show that

$$\delta' = -\kappa((\gamma \cdot T)N + (\gamma \cdot N)T).$$

Show that the second factor of $\delta'(t)$ is zero if and only if $\gamma(t)=0$, which says that, for this $t$, the curve $\gamma$ passes through the origin. Assuming that $\gamma(t)$ is never 0, deduce that $\delta$ is a regular curve except at those points corresponding to inflexions of $\gamma$. Fig. 2.19 shows the pedal curve of a limaçon – see 2.15(5), where we take $a=2.3$ and draw the pedal with respect to $(1, 0)$.

(2) **Parallels** Let $\gamma$ be a unit speed plane curve and let $d$ be a fixed real number. The curve $\delta$ defined by $\delta(t)=\gamma(t)+dN(t)$ is called the *parallel* to $\gamma$ at distance $d$. Show that $\delta$ is a regular curve except for values of $t$ where $\kappa(t)\neq0$ and $d=1/\kappa(t)$; also that for these irregular points $\delta(t)$ is the centre of curvature of $\gamma$ at $\gamma(t)$. See figs. 2.21 and 5.10.

(3) **Evolute** Given a unit speed plane curve $\gamma$ with $\kappa(t)$ never zero we can consider the locus of centres of curvature of $\gamma$, namely the curve

$$\varepsilon(t)=\gamma(t)+[1/\kappa(t)]N(t),$$

which is called the *evolute* of $\gamma$.

(i) Show that the evolute is a regular curve except for those values of $t$ where $\kappa'(t)=0$, i.e. for points of $\varepsilon$ corresponding to vertices of $\gamma$ (Compare 2.27(ii)).

(ii) Assuming that $\kappa'(t)<0$ on $I$, so that the radius of curvature $\rho$ is increasing on $I$, show that the arc-length on $\varepsilon$ from $t_0$ to $t_1>t_0$ is $\rho(t_1) - \rho(t_0)$.

(iii) With $\kappa'<0$ again, show the unit tangent and normal to $\varepsilon$ satisfy $T_\varepsilon=N$, $N_\varepsilon=-T$, and that the curvature of $\varepsilon$ is $\kappa_\varepsilon=-\kappa^3/\kappa'$ (i.e. $\rho_\varepsilon=\rho\rho'$). Hence the evolute has no inflexions.

(iv) Assume $\kappa>0$ on $I$. Deduce from (ii) that if a piece of string is wrapped round the evolute, one end being fastened at $\varepsilon(t_0)$ and the other

**Fig. 2.19.** Limaçon and its pedal curve

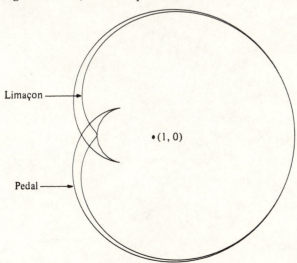

Limaçon

Pedal

$\bullet\,(1, 0)$

end starting at $\gamma(t_0)$, then as the string is unwrapped the ends are at $\varepsilon(t)$ and $\gamma(t)$ for all $t > t_0$. This says $\|\gamma(t) - \varepsilon(t)\| \overset{?}{=} \|\gamma(t_0) - \varepsilon(t_0)\| + $ (arc-length on $\varepsilon$ from $t_0$ to $t$.) See fig. 2.20.

Figure 2.21 shows some parallels and the evolute of a parabola: notice that the cusps on the parallels all lie on the evolute – compare (2) above – and that the point of the evolute corresponding to the minimum of the parabola, which is a vertex, is a cusp – compare (3) above.

(4) Let $r$ be a (differentiable) function of $t$ and let $X(t) = r(t) \cos t$, $Y(t) = r(t) \sin t$. Then $\gamma(t) = (X(t), Y(t))$ defines a plane curve by polar coordinates $(r, t)$. What is the condition for $\gamma$ to be regular? To be unit speed? Show that there are rather few functions $r$ which make the curve unit speed. Can you discover what the corresponding curves look like?

**Fig. 2.20.** String unwinding from the evolute

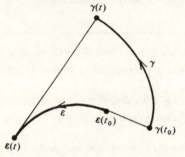

**Fig. 2.21.** Parallels and evolute of a parabola

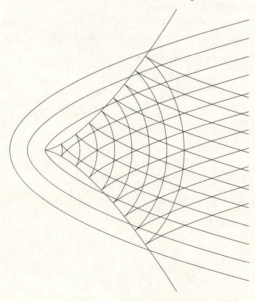

### 2.31 Parametrization by $x$ or $y$

Let $\gamma: I \to \mathbb{R}^2$, $\gamma(t) = (X(t), Y(t))$, be a (regular) curve in the plane. Suppose that $t_0 \in I$ and $X'(t_0) \neq 0$. Let $I_0$ be a small enough open interval containing $t_0$ so that, for all $t \in I_0$, $X'(t) \neq 0$. Let $X(I_0) = J$; then $X: I_0 \to J$ is a diffeomorphism. Using the change of parameter $X^{-1}$ applied to the part of $\gamma$ defined by $I_0$ we obtain

$$J \xrightarrow{X^{-1}} I_0 \xrightarrow{\gamma | I_0} \mathbb{R}^2$$

Calling this $\beta$, we have $\beta(x) = \gamma(X^{-1}(x)) = (X(X^{-1}(x)), Y(X^{-1}(x))) = (x, f(x))$ say where $f$ is the smooth function $Y \circ X^{-1}$. Thus the reparametrization $\beta$ of a part of $\gamma$ close to $t_0$ displays this part of $\gamma$ as the *graph of a function* $f$, with equation $y = f(x)$. We say that $\gamma$ *can be parametrized by* $x$ *close to any point where* $X'(t_0) \neq 0$, i.e. where the tangent line is not 'vertical' (fig. 2.22).

Similarly if $Y'(t_0) \neq 0$ there is a reparametrization $\beta$ of the form $\beta(y) = (g(y), y)$ for a smooth $g$, valid close to $y_0 = Y(t_0)$.

Note that, if $h$ is any reparametrization defined near $t_0$, for which $\gamma(h(u)) = (u, f(u))$ for some $f$, then $h_1^{-1} = X$, for $t$ close to $t_0$, so $f = Y \circ X^{-1}$. Thus the $f$ is uniquely determined by $\gamma$ and $t_0$ – we should hardly expect the image of $\gamma$ for $t$ close to $t_0$ to be the graph of two *different* functions. Beware however, that we could have $\gamma(t_0) = \gamma(t_1)$ say, and then the pieces of the image of $\gamma$ given by $t$ close to $t_0$ and $t$ close to $t_1$ may well be graphs of different functions.

### 2.32 Exercise

At $(0, 0)$ and $(1, 0)$ the curve $y^2 = x - x^3$ in $\mathbb{R}^2$ has vertical tangent. Assuming that there exists a smooth function $X(t)$, defined near $t = 0$, such that $t^2 = X(t) - (X(t))^3$ (so that $(X(t), t)$ parametrizes the curve near $(0, 0)$), find the curvature at the point with parameter $t = 0$. Similarly for a parametrization of the same form near $(1, 0)$. Comparing with 2.26(3), are your answers

**Fig. 2.22.** Parametrization by $x$

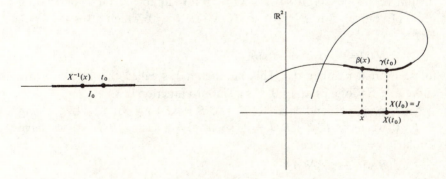

the limiting values of the $\kappa(t)$ there as $t \to 0, 1$? What about the sign of $\kappa$? (The assumption is true; compare 4.14.)

## Space curves

We shall now take off from the plane into space and consider *space curves* $\gamma: I \to \mathbb{R}^3$. By 2.22 any (regular) space curve can be made unit speed by a mere change of parameter, so *we shall assume all our curves are unit speed* below. The necessary changes in the formulas for the general case are easy to supply.

### 2.33     Serret–Frenet formulae for unit speed space curves

Let $\alpha: I \to \mathbb{R}^3$ be unit speed, so that $\alpha'(s) = T(s)$, the unit tangent vector (2.5). Also the curvature is given by

$$\kappa(s) = \| T'(s) \|$$

and provided $\kappa(s) \neq 0$ there is a unit principal normal $N(s)$, perpendicular to $T(s)$, with (see 2.23)

$$T'(s) = \kappa(s)N(s).$$

*We shall continue to assume* $\kappa(s)$ is non-zero in what follows. This is not as artificial as it sounds, for in fact 'most' space curves have nowhere zero curvature. Compare 9.9.

Since $T(s)$, $N(s)$ are perpendicular unit vectors there is a unique unit vector $B(s)$ called the *binormal vector* perpendicular to both and such that $T(s)$, $N(s)$, $B(s)$ is right-handed (i.e. if the components are written as the successive rows of a $3 \times 3$ matrix, then the determinant is $+1$ rather than $-1$). The plane through $\alpha(s)$ spanned by $N(s)$, $B(s)$ is called the *normal plane* at $s$ (or more loosely, at $\alpha(s)$) and the plane spanned by $T(s)$, $N(s)$ is called the *osculating plane* there. The latter has higher contact with $\alpha$ than do other planes; see 2.35 below.

For the purpose of calculation let us drop the $s$. We know $N'$ is perpendicular to $N$ (see 2.6(3)), so $N' = \lambda T + \tau B$ for suitable $\lambda$, $\tau \in \mathbb{R}$ (depending on $s$, of course). But $T \cdot N = 0$ so $T' \cdot N + T \cdot N' = 0$, which gives $\kappa + \lambda = 0$. Hence

$$N'(s) = -\kappa(s)T(s) + \tau(s)B(s),$$

for some real number $\tau(s)$. This number $\tau(s)$ is called the *torsion* of the curve at $s$. So long as $\kappa(s) \neq 0$, $\tau$ is a smooth function of $s$.

Next, $B'$ is perpendicular to $B$, say $B' = \mu T + \nu N$ for $\mu$, $\nu \in \mathbb{R}$. Using $B \cdot T = 0$ we have $B' \cdot T + B \cdot T' = 0$, so $\mu + 0 = 0$. Using $B \cdot N = 0$ we have similarly $\nu = -\tau$. Hence

$$B'(s) = -\tau(s)N(s).$$

These three formulae, for $T'$, $N'$, $B'$, are the formulae we need.

**2.34    Distance-squared functions on a unit speed space curve**

Let $\alpha$: $I \to \mathbb{R}^3$ be unit speed, with $\kappa(s)$ never zero, and let $u \in \mathbb{R}^3$. Define $f$: $I \to \mathbb{R}$ by $f(s) = \|\alpha(s) - u\|^2 = (\alpha(s) - u) \cdot (\alpha(s) - u)$. Dropping most of the $s$'s we have

$$\tfrac{1}{2}f'(s) = (\alpha - u) \cdot T$$
$$\tfrac{1}{2}f''(s) = 1 + (\alpha - u) \cdot \kappa N$$
$$\tfrac{1}{2}f'''(s) = (\alpha - u) \cdot \kappa' N + (\alpha - u) \cdot \kappa (-\kappa T + \tau B).$$

Hence

$$f'(s) = 0 \Leftrightarrow u \text{ is in the normal plane at } s$$
$$f'(s) = f''(s) = 0 \Leftrightarrow u = \alpha + \kappa^{-1} N + \mu B \text{ (at } s) \text{ for some } \mu \in \mathbb{R}$$
$$f'(s) = f''(s) = f'''(s) = 0 \Leftrightarrow \begin{cases} \tau \neq 0 \text{ and } u = \alpha + N/\kappa - (\kappa'/\kappa^2 \tau)B \text{ (at } s) \\ \text{or } \tau = \kappa' = 0 \text{ (at } s) \text{ and } u = N/\kappa + \mu B. \end{cases}$$

Now the distance-squared function from $u$ measures contact with the sphere, centre $u$, through a given point of the curve, since it differs by a constant from the function $g$ of 2.9, where $F$ is given by 2.7. Compare the plane case, 2.13. Thus the last equation above for $u$ shows that, provided $\kappa(s) \neq 0$ and $\tau(s) \neq 0$, there is a *unique* sphere having at least 4-point contact with $\alpha$ for the parameter value $s$. This sphere is called the *sphere of curvature* at $s$ and its *centre $u$ the centre of spherical curvature* at $s$.

**2.35    Height functions on unit speed space curves.**

With $\alpha$ as in 2.34, and $u$ a unit vector in $\mathbb{R}^3$, define $f$: $I \to \mathbb{R}$ by $f(s) = \alpha(s) \cdot u$. (Thus $f(s)$ is the distance of $\alpha(s)$ from the plane through the origin perpendicular to $u$.) We find:

$$f'(s) = 0 \qquad\qquad \Leftrightarrow u \text{ is a vector in the normal plane at } s$$
$$f'(s) = f''(s) = 0 \qquad \Leftrightarrow u = \pm B(s)$$
$$f'(s) = f''(s) = f'''(s) = 0 \Leftrightarrow u = \pm B(s), \tau(s) = 0$$
$$f'(s) = \cdots = f^{(4)}(s) = 0 \Leftrightarrow u = \pm B(s), \tau(s) = \tau'(s) = 0.$$

The verification of these is left as an exercise. It follows that the osculating plane at $s$ is the unique plane having at least 3-point contact with $\alpha$ for that value of $s$, and that it has higher contact if $\tau(s) = 0$.

**2.36    Exercises**

(1) The *vector product* of $a = (a_1, a_2, a_3)$ and $b = (b_1, b_2, b_3) \in \mathbb{R}^3$ is $a \times b$ $= (a_2 b_3 - a_3 b_2, a_3 b_1 - a_1 b_3, a_1 b_2 - a_2 b_1)$. Show that $B(s) = T(s) \times N(s)$ (in the notation of 2.33).

(2) For the unit speed helix $\alpha(s) = (1\sqrt{2})(\cos s, \sin s, s)$, show that $\kappa(s) = 1/\sqrt{2}$ for all $s$. Find expressions for $T(s)$, $N(s)$ and show $B(s) = (1/\sqrt{2})(\sin s, -\cos s, 1)$ for all $s$. Find $\tau(s)$. Find the centre of spherical curvature at $s$.

(3) Let $f$ be the height function of 2.35. Show that $\kappa(s)=0$ if and only if $f'(s)=f''(s)=0$ for all $u$ in the normal plane at $s$, i.e. all $u$ satisfying $T(s)\cdot u=0$.

(4) Let $\alpha\colon I\to\mathbb{R}^3$ be unit speed, with $\alpha(0)=(0,0,0)$, $T(0)=(1,0,0)$, $N(0)=(0,1,0)$. Show that $B(0)=(0,0,1)$ and

$\alpha''(0)=(0,\kappa,0)$ ($\kappa$, $\tau$ etc. are all at $s=0$)

$\alpha'''(0)=(-\kappa^2,\kappa',\kappa\tau)$

$\alpha^{(4)}(0)=(-3\kappa\kappa',\kappa''-\kappa^3-\kappa\tau^2,2\kappa'\tau+\kappa\tau')$.

(5) Let $\alpha\colon I\to\mathbb{R}^3$ be unit speed, and $s_0\in I$. Define $\gamma\colon I\to\mathbb{R}^2$ by $\gamma(s)=(\alpha(s)\cdot T(s_0),\alpha(s)\cdot N(s_0))$. (This is the plane curve obtained by projecting $\alpha$ to the osculating plane at $s_0$. Note that $\gamma$ need not be unit speed.) Show that, restricting $\gamma$ to a suitably small open interval containing $s_0$, it is a regular curve and that its curvature at $s_0$ equals the curvature of $\alpha$ at $s_0$. Try projecting to the planes spanned by $T(s_0)$, $B(s_0)$ or by $N(s_0)$, $B(s_0)$.

(6) Show, after the manner of 2.28(5), that curvature and torsion of space curves are unaffected by rotations and translations of $\mathbb{R}^3$.

(7) Show that the difference between the length of a small arc of a space curve and the length of the corresponding chord is of order at least $s^3$ (parametrizing by arc-length $s$). (One expects at least $s^2$.)

(8) Let $\alpha$ be a unit speed space curve; write

$\alpha^{(n)}(s)=a_n(s)T(s)+b_n(s)N(s)+c_n(s)B(s)$

for smooth functions $a$, $b$, $c$. Prove the reduction formulae

$a_{n+1}=a'_n-\kappa b_n,$

$b_{n+1}=b'_n+\kappa a_n-\tau c_n,$

$c_{n+1}=c'_n+\tau b_n.$

**2.37     Project**

There is a topic dear to the hearts of most authors of books on curves which we have not mentioned at all. To what extent does the curvature function *determine* a plane curve, or the curvature and torsion functions determine a space curve? Since the curvature of a unit speed plane curve $\alpha$ has to do with the second derivative $\alpha''$ we should expect that $\alpha$ is fixed by the curvature once we are given a point $\alpha(t_0)$ and the tangent $\alpha'(t_0)$ there. The same conclusion is suggested by the interpretation $\kappa(s)=d\psi/ds$ (see 2.24) where $\psi$ is the tangent angle. More geometrically we have $\delta\psi=\kappa\delta s$ so one expects that, starting with a point and a tangent line, the curve can be constructed 'step by step' from the function $\kappa$ (fig. 2.23.). It is important to realize that $\kappa$ must be given as a function of arc-length – or, to put it another way, given a function $\kappa\colon I\to\mathbb{R}$, the curve constructed from $\kappa$ will have unit speed.

Prove for yourself, or look up in for example Thorpe (1979) or DoCarmo (1976) the following facts.

(1) Let $\alpha_1\colon I\to\mathbb{R}^2$ and $\alpha_2\colon I\to\mathbb{R}^2$ be unit speed, with $\alpha_1(t_0)=\alpha_2(t_0)$, $\alpha'_1(t_0)=\alpha'_2(t_0)$ for some $t_0\in I$ and $\kappa_1(t)=\kappa_2(t)$ for all $t\in I$ ($\kappa_i=$curvature of $\alpha_i$). Then $\alpha_1(t)=\alpha_2(t)$ for all $t\in I$.

**Fig. 2.23.** Construction of the curve from $\kappa$

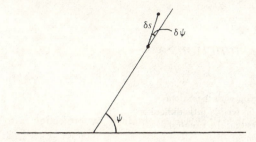

(2) Let $\alpha_1 \colon I \to \mathbb{R}^3$ and $\alpha_2 \colon I \to \mathbb{R}^3$ be unit speed, with $\alpha_1(t_0) = \alpha_2(t_0)$, $\alpha_1'(t_0) = \alpha_2'(t_0)$, $\alpha_1''(t_0) = \alpha_2''(t_0) \neq 0$ for some $t \in I$ and $\kappa_1(t) = \kappa_2(t)$, $\tau_1(t) = \tau_2(t)$ for all $t \in I$. Then $\alpha_1(t) = \alpha_2(t)$ for all $t \in I$.

What do these tell you about a plane curve with $\kappa(t) = 0$ for all $t$? With $\kappa(t)$ constant? A space curve with $\tau(t) = 0$ for all $t$?

# 3

## More about functions

'I fear that I bore you with these details,
but I have to let you see my little difficulties,
if you are to understand the situation.'
(*A Scandal in Bohemia*)

Start with a function defined on a curve, as in chapter 2, say the height function $f(t)$ defined on an ellipse in some chosen direction $a$. What happens if we reparametrize the ellipse, or move the origin in the plane, or both? The effect is to change $f(t)$ into the function $g(t) = f(h(t)) + c$, where $h$ is the change of parameter and $c$ is a constant. This is a different function, of course, but it has a lot in common with the old function – for example the number of derivatives of $g$ vanishing at any point $t_0$ is the same as the number of derivatives of $f$ vanishing at $h(t_0)$.

In this chapter we make the change from $f$ to $g$ into an equivalence relation – so-called right-equivalence (because in $f(h(t))$, $h$ is to the right of $f$!) In the above, $g$ at $t_0$ is right-equivalent to $f$ at $h(t_0)$.

Amazingly, it is usually possible to tell very easily when two functions (at appropriate points) are right-equivalent. Furthermore, 'most' functions are right-equivalent, at a point, to one of the functions $\pm t^k$, at $t = 0$, where $k$ is an integer $\geqslant 0$. (The exceptional functions are called 'flat'.) It is great good luck to have so simple a 'normal form' to which functions can be reduced, and the source of this luck is the fact that we are working here only with functions of one variable $t$. With more than one variable both 'classification' (listing equivalence classes) and 'recognition' (telling to which equivalence class a given function belongs) are central and difficult problems. We say a little about these in chapter 11.

### Right equivalence

Let $U_i$, $i = 1$, 2, be open subsets of the real line $\mathbb{R}$, let $t_i$ be a point of $U_i$ and $f_i: U_i \to \mathbb{R}$ be smooth functions. Thus $f_i$ is defined at any rate on a neighbourhood of $t_i$.

### 3.1 Definition

We say that $f_1$ (at $t_1$) and $f_2$ (at $t_2$) are *right-equivalent* (written $\mathscr{R}$-equivalent) if there exist open intervals $V_i \subset U_i$, with $t_i \in V_i$, a diffeomorphism $h \colon V_1 \to V_2$ and a constant $c \in \mathbb{R}$ such that

$$h(t_1) = t_2 \quad \text{and} \quad f_1(t) = f_2(h(t)) + c$$

for all $t \in V_1$.

Thus $f_1$, close to $t_1$, is obtained from $f_2$, close to $t_2$, by changing the parameter and adding a constant. Note that $c = f_1(t_1) - f_2(t_2)$. The definition can also be expressed by a *commutative diagram*

$$U_1 \supset V_1 \xrightarrow{f_1} \mathbb{R}$$
$$h\downarrow \qquad \quad \downarrow$$
$$U_2 \supset V_2 \xrightarrow{f_2} \mathbb{R}$$

where the right-hand vertical arrow is $x \mapsto x - c$.

It is not difficult to check that $\mathscr{R}$-equivalence is an equivalence relation in the following sense: (*i*) $f$ (at $t_1$) is $\mathscr{R}$-equivalent to $f$ (at $t_1$); (*ii*) if $f_1$ (at $t_1$) is $\mathscr{R}$-equivalent to $f_2$ (at $t_2$) then the reverse holds; (*iii*) if $f_1$ (at $t_1$) is $\mathscr{R}$-equivalent to $f_2$ (at $t_2$), and the latter is $\mathscr{R}$-equivalent to $f_3$ (at $t_3$) then $f_1$ (at $t_1$) is $\mathscr{R}$-equivalent to $f_3$ (at $t_3$).

### 3.2 Examples

(1) $f_1(t) = t^2$ and $f_2(t) = -t^2$ are not $\mathscr{R}$-equivalent (taking $t_1 = t_2 = 0$). For if $h$ as above existed then (since $c = 0$) $t^2 = -(h(t))^2$ for all $t$ close to 0, but this equation clearly implies that $t = h(t) = 0$. (The same goes for $t^{2n}$ and $-t^{2n}$.) On the other hand with say $t_1 = t_2 = 1$, $f_1$ is $\mathscr{R}$-equivalent to $f_2$ by taking $h(t) = (2 - t^2)^{\frac{1}{2}}$ and $c = 2$. A change of parameter, plus a constant, will not turn the minimum of $t^2$ at 0 into the maximum of $-t^2$ at 0 but it will turn the increasing function $t^2$ at 1 into the decreasing function $-t^2$ at 1.

(2) A function defined on a parametrized curve $\gamma \colon I \to \mathbb{R}$ and the function obtained by any change of parameter $h$ as in 2.16 are certainly $\mathscr{R}$-equivalent for any $t_1 \in I$, taking $t_2 = h(t_1)$.

We shall use the following notation. Let $t_0 \in \mathbb{R}$. Then $f \colon \mathbb{R}, t_0 \to \mathbb{R}$ denotes a function defined on *some* neighbourhood of $t_0$. We shall regard two such functions as equal if they coincide on some (possibly smaller) neighbourhood of $t_0$. (This leads to the formal idea of a *germ*, but we shall not need the formality. See for example Bröcker and Lander (1975).) When $f_i \colon \mathbb{R}$, $t_i \to \mathbb{R}$, $i = 1, 2$ and we say that $f_1$ and $f_2$ are $\mathscr{R}$-equivalent this always means that $f_1$ (at $t_1$) is $\mathscr{R}$-equivalent to $f_2$ (at $t_2$). The notation $f \colon \mathbb{R}, t_0 \to \mathbb{R}, c$ means, in addition, that $f(t_0) = c$.

### 3.3     Theorem

Let $f\colon \mathbb{R},\, t_0 \to \mathbb{R}$ be smooth, and let $k \geqslant 0$. Suppose that $f^{(p)}(t_0)=0$ for all $p$ with $1 \leqslant p \leqslant k$, while $f^{(k+1)}(t_0) \neq 0$. Then $f$ is $\mathscr{R}$-equivalent to $g$: $\mathbb{R},\, 0 \to \mathbb{R}$ defined by $g(t)= \pm t^{k+1}$ where we have $+$ or $-$ according as $f^{(k+1)}(t_0)$ is $>0$ or $<0$.

### 3.4     Hadamard's Lemma

Let $f\colon \mathbb{R},\, t_0 \to \mathbb{R}$ be smooth, and suppose $f^{(p)}(t_0)=0$ for all $p$ with $1 \leqslant p \leqslant k$. Then there is a smooth function $f_1\colon \mathbb{R},\, t_0 \to \mathbb{R}$ such that $f(t)=f(t_0)+ (t-t_0)^{k+1}f_1(t)$ for all $t$ in some neighbourhood of $t_0$. Further, if $f^{(k+1)}(t_0) \neq 0$ then $f_1(t_0) \neq 0$. (When $k=0$, there are no such $p$ and the result always holds.)

The theorem is an easy consequence of the lemma, as follows. Define

$$h(t)=(t-t_0)(\pm f_1(t))^{1/(k+1)}$$

where the sign $\pm$ is that of $f_1(t_0)$. Then $h(t_0)=0$, $h'(t_0)>0$ so that $h$ is a diffeomorphism on *some* neighbourhood of $t_0$. Further $g(h(t)) =(t-t_0)^{k+1}f_1(t)=f(t)-f(t_0)$ where $g$ has the form in the theorem. It is easy to check that the signs of $f_1(t_0)$ and $f^{(k+1)}(t_0)$ coincide (indeed $f^{(k+1)}(t_0) =(k+1)!f_1(t_0)$, using Leibniz' rule on the formula of the lemma). Hence $f$ and $g$ are indeed $\mathscr{R}$-equivalent as claimed. $\qquad\square$

***Proof of lemma***     It is enough to prove the lemma in the special case $t_0=f(t_0)=0$. Thus suppose $F\colon \mathbb{R},\, 0 \to \mathbb{R},\, 0$ is smooth and $F^{(p)}(0)=0$ for $1 \leqslant p \leqslant k$. We prove that $F(t)=t^{k+1}F_1(t)$ for a smooth $F_1$ and all $t$ close to 0. The general case follows by $F(t)=f(t+t_0)-f(t_0)$, $f_1(t)=F_1(t-t_0)$. Note that $f_1(t_0) \neq 0$ follows from $f^{(k+1)}(t_0) \neq 0$ by repeated differentiation.

The proof is by induction on $k$. For $k=0$ we have $F(0)=0$ and require $F(t)=tF_1(t)$. Now

$$\int_0^1 \frac{\mathrm{d}}{\mathrm{d}u}\, F(tu)\,\mathrm{d}u = [F(tu)]_0^1 = F(t)-F(0)=F(t).$$

Hence

$$F(t)= \int_0^1 \frac{\mathrm{d}}{\mathrm{d}u}\, F(tu)\,\mathrm{d}u = \int_0^1 tF'(tu)\,\mathrm{d}u = t\int_0^1 F'(tu)\,\mathrm{d}u.$$

Thus we can take $F_1(t)=\int_0^1 F'(tu)\,\mathrm{d}u$, which is a smooth function – its derivatives are given by 'differentiation under the integral sign',

$$F_1'(t)= \int_0^1 \frac{\mathrm{d}}{\mathrm{d}t}\, (F'(tu))\,\mathrm{d}u,\ \text{etc.}$$

The induction step is easier. Assume the result for $k$ $(k \geqslant 0)$ and suppose $F^{(p)}(0)=0$ for $1 \leqslant p \leqslant k+1$. Then by the case $k=0$, $F(t)=tF_2(t)$ say and this gives $F^{(p)}(t)=tF_2^{(p)}(t)+pF_2^{(p-1)}(t)$ by Leibniz' rule. Hence $F_2^{(p)}(0)=0$ for $1 \leqslant p \leqslant k$. Using the induction hypothesis, $F_2(t)=t^{k+1}F_1(t)$ for a smooth $F_1$, giving $F(t)=t^{k+2}F_1(t)$ as required. $\qquad\square$

### 3.5    Remarks

(1) If $k$ is *even* then $t^{k+1}$ and $-t^{k+1}$ (both at $t=0$) are $\mathscr{R}$-equivalent by $h(t)=-t$. As noted above (3.2(1)) for $k$ odd they are *not* $\mathscr{R}$-equivalent.

(2) For $k \neq l$, neither of $\pm t^{k+1}$ is $\mathscr{R}$-equivalent to $t^{l+1}$ (both at $t=0$). For otherwise we could find $h: \mathbb{R}, 0 \to \mathbb{R}$ with $h(0)=0$ and $\pm (h(t))^{k+1} = t^{l+1}$. With $k > l$ we differentiate both sides $l+1$ times and put $t=0$: the left side becomes 0 and the right side $\neq 0$, giving a contradiction.

(3) There are many-variable versions of the Hadamard lemma, which are not much harder to prove. For instance, suppose that $f: U \to \mathbb{R}$ where $U$ is a neighbourhood of $(u_1, u_2)$ in $\mathbb{R}^2$, and $f$ is 'smooth', i.e. all partial derivatives of all orders exist and are continuous. Suppose $f(u_1, x_2)=0$ for all $x_2$ in a neighbourhood of $u_2$; then there exists a smooth $f_1$ so that $f(x_1, x_2)=(x_1 - u_1) f_1(x_1, x_2)$ for all $(x_1, x_2)$ near $(u_1, u_2)$.

### 3.6    Definition

Suppose that $f: \mathbb{R}, t_0 \to \mathbb{R}$ is $\mathscr{R}$-equivalent to $\pm t^{k+1}$. (The necessary and sufficient condition for this is $f^{(p)}(t_0)=0$ for all $p$ with $1 \leq p \leq k$, and $f^{(k+1)}(t_0) \neq 0$, by 3.1 and 3.3.) Then, for $k \geq 0$, we say that $f$ has *type $A_k$* at $t_0$, or an $A_k$ *singularity* at $t_0$. Thus type $A_0$ just means $f'(t_0) \neq 0$. We also say that $f$ has type $A_{\geq k}$ when $f^{(p)}(t_0)=0$ for $1 \leq p \leq k$. So either $f$ has type $A_l$ for some $l \geq k$ or *all* derivatives of $f$ vanish at $t_0$ (compare 3.9).

### 3.7    Remark

This has the following geometrical interpretation. The function $f$ has type $A_k$ $(k \geq 0)$ at $t_0$ iff the graph of $f$ (given by $\gamma(t)=(t, f(t))$) has $(k+1)$-point contact with the line in $\mathbb{R}^2$ parallel to the $t$-axis, through $(t_0, f(t_0))$. (Compare 2.12(2) and fig. 2.9.) We could extend this to define two graphs $y=f(t)$ and $y=g(t)$ in the $(t, y)$-plane to have contact of order $(k+1)$ at an intersection point $(t_0, y_0)$, where $y_0=f(t_0)=g(t_0)$, precisely when $f-g$ has an $A_k$ singularity at $t_0$. This can be reconciled with the notion of contact given in chapter 2 – see 4.27.

It has to be admitted that the above definition could have been framed without ever defining $\mathscr{R}$-equivalence or proving the theorem 3.3. What the definition of $\mathscr{R}$-equivalence and theorem 3.3 do is to make explicit the equivalence relation for functions on curves corresponding with change of parameter, and to list the equivalence classes in a simple 'normal' form. Another way of thinking of theorem 3.3 is to say that, if $f(t) = \pm t^{k+1} + \phi(t)$ ($t$ close to 0) where $\phi(t)$ consists of 'higher terms', then a mere change of parameter will eliminate $\phi(t)$ altogether. Here 'higher terms' means that $\phi^{(p)}(0)=0$ for $1 \leq p \leq k+1$, or, by the Hadamard lemma 3.4 that $\phi(t) = t^{k+2} \phi_1(t)$ for a smooth $\phi_1$. In this sense $\pm t^{k+1}$ is said to be '$(k+1)$-sufficient' or '$(k+1)$-determined'. Knowing when, in some situation, higher terms can be ignored, is of crucial importance in applications of mathematics.

It is worth mentioning that the corresponding problem for functions of two (or more) variables is *very* much harder. Much work has been done on the classification, with various equivalence relations suited to various applications, and long lists of normal forms exist. Some of the longest are due to the Russian school led by V. I. Arnold. See for example Arnold (1981).

**3.8    Exercises**

(1) Use the Hadamard lemma 3.4 to prove the following version of Taylor's theorem: let $f: \mathbb{R}, t_0 \rightarrow \mathbb{R}$ be smooth, write $a_p$ for $f^{(p)}(t_0)/p!$ and let $k$ be an integer $\geq 0$. Then there exists a smooth function $f_1: \mathbb{R}, t_0 \rightarrow \mathbb{R}$ such that

$$f(t) = \sum_{p=0}^{k} a_p(t - t_0)^p + (t - t_0)^{k+1} f_1(t)$$

for all $t$ in a neighbourhood of $t_0$. (Just apply the lemma to $f(t) - \Sigma_0^k$.) This can also be written

$$f(t + t_0) = \sum_{p=0}^{k} a_p t^p + t^{k+1} f_2(t)$$

for all $t$ in a neighbourhood of 0, where $f_2(t) = f_1(t + t_0)$.

(2) Find explicitly the number $k$ and the function $f_1$ of lemma 3.4 in the case $f(t) = t^2 + t^3$ where (i) $t_0 = 0$, (ii) $t_0 = -1$. (No integration is necessary.) Hence find explicitly the change of variable reducing $f$ to $\pm t^{k+1}$ as in the lines following the statement of 3.4.

(3) Suppose that $f: \mathbb{R}, 0 \rightarrow \mathbb{R}$ is smooth and $f(0) = 0$. Define

$$f_1(t) = \begin{cases} f(t)/t & t \neq 0 \\ f'(0) & t = 0. \end{cases}$$

(This is one way of explicitly writing down $f_1$ with $f(t) = t f_1(t)$.) Show that $f'_1(0)$ exists and equals $\frac{1}{2} f''(0)$. (Hint: use the definition of $f'_1(0)$ to show that it equals

$$\lim_{t \to 0} \frac{f(t) - t f'(0)}{t^2}$$

and apply l'Hôpital's rule.)

Show that $f_1^{(k)}(0)$ exists and equals $f^{(k+1)}(0)/(k+1)$. Since $f_1$ is clearly smooth away from 0 this shows directly that $f_1$ is smooth in a neighbourhood of 0 and establishes the case $k = 1$ of the Hadamard lemma. (This is a bit harder. Here is one suggestion. First prove by induction that

$$f_1^{(k)}(t) = 1/t^{k+1} \int_0^t t^k f^{(k+1)}(t) dt, \quad \text{for} \quad t \neq 0.$$

Then prove the required result by induction, writing down the definition of $f_1^{(k+1)}(0)$, using the above formula and applying L'Hôpital's rule.)

(4) Let $\gamma: \mathbb{R}, 0 \to \mathbb{R}^2$ be $\gamma(t) = (t, a_2 t^2 + a_3 t^3 + a_4 t^4 + a_5 t^5) = (t, Y(t))$ say, where the $a_i$ are constants. Let $f(t) = (a - t)^2 + (b - Y(t))^2$ be the distance-squared function on the curve $\gamma$ from $(a, b) \in \mathbb{R}^2$. For each $k \geqslant 0$ find the conditions on $a, b, a_1, \ldots, a_5$ that $f$ should have an $A_k$ singularity at 0. Show that $k > 5$ is impossible. (What do you think the highest possible $k$ is when $Y$ goes up to $a_n t^n$?)

(5) (Easy.) Repeat (4) for the height function on $\gamma$ in the direction $u = (a, b)$. (Thus $f(t) = \gamma(t) \cdot u$.)

(6) Let $\gamma: \mathbb{R}, t_0 \to \mathbb{R}^2$ be a unit speed plane curve with $\kappa(t)$ never zero. Let $u \in \mathbb{R}^2$ and $f(t) = (\gamma(t) - u) \cdot (\gamma(t) - u)$. Define $g_1(t) = (\gamma(t) - u) \cdot N(t)$ (distance to tangent) and $g_2(t) = (\gamma(t) - u) \cdot T(t)$ (distance to normal). Show that $g_1'(t) = -\kappa(t) f'(t)$ and deduce $f$ is of type $A_k$ at $t_0$ iff $g_1$ is of type $A_k$ at $t_0$. Show that for $k \geqslant 0$ $f$ is of type $A_{k+1}$ at $t_0$ iff $g_2$ is of type $A_k$ at $t_0$ and $g_2(t_0) = 0$.

(7) Let $f: \mathbb{R}, 0 \to \mathbb{R}, 0$ be smooth and have type $A_k$ $(k \geqslant 1)$ at 0. The 'Jacobian ideal' of $f$ at 0 is the set of smooth functions $\mathbb{R}, 0 \to \mathbb{R}, 0$ of the form $g(t) = f'(t) f_1(t)$ where $f_1: \mathbb{R}, 0 \to \mathbb{R}$ is smooth. Show that the Jacobian ideal consists precisely of functions of the form $t \mapsto t^k g_1(t)$ where $g_1: \mathbb{R}, 0 \to \mathbb{R}$ is smooth. (This uses the Hadamard lemma 3.4 with $t_0 = 0$. For proving the result one way round you will need to use the fact that $1/f_2(t)$ is smooth near 0 provided $f_2$ is smooth near 0 and $f_2(0) \neq 0$.)

## Flat functions

Are *all* functions of type $A_k$ for some $k$? The definition requires that some $f^{(p)}(t_0)$ should be nonzero. For an *analytic* function $f$, defined by a convergent power series,

$$f(t) = \sum_{p \geqslant 0} a_p (t - t_0)^p,$$

where the $a_p$ are real numbers, we simply take the smallest $p$ for which $a_p$ is nonzero, and $f$ then has type $A_{p-1}$ at $t_0$. If no such $p$ exists then $f$ is identically zero so that is the only exception. But not all functions are analytic, as the following example shows.

**3.9      Example**
    Let $f(t) = \exp(-1/t)$ for $t > 0$ and $f(t) = 0$ for $t \leqslant 0$. Then $f^{(p)}(0) = 0$ for all $p \geqslant 0$, so that the power series expansion of $f$ is identically zero, and consequently does not equal $f$, which is plainly not identically zero. Of course $f$ fails to be of type $A_k$ for every $k$, at 0, but $f$ is smooth at every point of $\mathbb{R}$ (either directly from the formula or from $f^{((p)}(0) = 0$ for all $p$).

To prove that all derivatives of $f$ vanish at 0 we argue by induction. We are given the case $p=0$. Now

$$f^{(p+1)}(0) = \lim_{t \to 0} \frac{f^{(p)}(t) - f^{(p)}(0)}{t} = \lim_{t \to 0} \frac{f^{(p)}(t)}{t}$$

by the induction hypothesis. We need only consider $t > 0$, and clearly $f^{(p)}(t)/t = P_p(1/t)e^{-1/t}$ where $P_p$ is a polynomial, of degree say $n$. Then $t^{2n} P_p(1/t) \to 0$ as $t \to 0$, so that $f^{(p)}(t)/t = t^{2n} P_p(1/t)(e^{-1/t}/t^{2n})$ and the final factor $\to 0$ as $t \to 0$ (for example by repeated application of L'Hôpital's invaluable rule, but it's best to put $u = 1/t$ and consider $u \to \infty$). This proves the induction step and hence the result. $\qquad\square$

The function of 3.9 is called *flat* at 0, this word meaning that all derivatives $f^{(p)}(t_0)$, $p \geqslant 1$, vanish at $t_0$. Any flat function slips through the $A_k$-net: it is not of type $A_k$ for any $k$. Flat functions, and their close relatives bump functions, are a vital tool in the theory and applications of smooth functions: they are not at all the unnecessary nuisance they appear to be. For more information see Bröcker and Lander (1975). Incidentally since a flat function has to satisfy 'infinitely many conditions' $f^{(p)}(t_0) = 0$ one might expect them to be 'infinitely rare', and in a sense they are.

### 3.10    Exercises

(1) With $f$ as in 3.9, let

$$g(t) = \frac{f(t)}{f(t) + f(\varepsilon - t)}$$

where $\varepsilon > 0$. Show that $g$ is smooth on $\mathbb{R}$ and that $g(t) = 0$ for $t \leqslant 0$, $g(t) = 1$ for $t \geqslant \varepsilon$. This $g$ is called a *bump function*.

(2) Show that *any* power series can arise as the Taylor series of a smooth function $f: \mathbb{R}, 0 \to \mathbb{R}$ at the origin. (Hint: let $\delta_n > 0$ and let $\phi_n$ be a smooth function $\mathbb{R} \to \mathbb{R}$ with $\phi_n(t) = 0$ if $|t| > \delta_n$ and $\phi_n(t) = 1$ if $|t| < \frac{1}{2}\delta_n$ (compare (1)). Now let $\sum a_n t^n$ be any power series and consider the sum $\sum_{n=0}^{\infty} a_n \phi_n(t) t^n$. Choose $\delta_n > 0$ so that $|a_n \phi_n(t) t^n| \leqslant (\frac{1}{2})^n$ for all $t \in \mathbb{R}$. Now use standard theorems on uniform convergence.)

### Jets

We conclude this chapter with a brief discussion of jets. It is useful in applications to have a finite-dimensional approximation to the (very large) space of smooth functions, the approximation to a function picking out at any rate some of its features. One way to do this locally is to approximate a function, defined near $t_0$, by part of its Taylor series at $t_0$ (of course this is a dismal failure with flat functions). The Taylor series at

$t_0$ can be written in the form

$$f(t_0)+tf'(t_0)+\frac{t^2}{2!}f''(t_0)+\frac{t^3}{3!}f'''(t_0)+\cdots$$

where we expand $f(t+t_0)$ rather than $f(t)$, so that $t$ is close to 0 rather than to $t_0$ (and we also avoid equating the series to $f(t+t_0)$, of course!).

### 3.11 Definition

Let $k\geqslant 1$ be an integer. The *k-jet of f at* $t_0$ is the polynomial

$$j^k f(t_0)=tf'(t_0)+\frac{t^2}{2!}f''(t_0)+\cdots+\frac{1}{k!}t^k f^{(k)}(t_0)$$

obtained by truncating the Taylor series to degree $k$ and deleting the constant term. Two $k$-jets are called *equal* when they are identically the same as polynomials. Sometimes we want to include the constant term in a jet. We refer to $f(t_0)+j^k f(t_0)$ as the *k-jet with constant* of $f$ at $t_0$.

### 3.12 Examples

(1) Clearly $f$ has type $A_k$ at $t_0$ iff $j^k f(t_0)=0$ and $j^{k+1}f(t_0)\neq 0$ (strictly this assumes $k\geqslant 1$, but thinking of $j^0 f(t_0)$ as 0 this works for $k=0$ too!). Also $f$ has type $A_m$ for some $m\geqslant k$ (type $A_{\geqslant k}$, we could say), iff $j^k f(t_0)=0$ and *some* derivative of $f$ is nonzero at $t_0$.

(2) Let $f(t)=(1+t)\sin(t^2)$. Then $j^3 f(0)=t^2+t^3$ since the Taylor series of $\sin(t^2)$ at 0 starts off $t^2-\frac{1}{6}t^6$.

(3) Let $f(t)=3+2t-t^3$. Then $j^k f(1)=-t+\frac{1}{2}(-6)t^2+\frac{1}{6}(-6)t^3=-t-3t^2-t^3$, for any $k\geqslant 3$. Of course we can work this out by evaluating $f(t+1)$ directly, and deleting the constant term.

(4) The $k$-jet $j^k(f+g)(t_0)$ is clearly equal to $j^k f(t_0)+j^k g(t_0)$. Since the Taylor series of a product is the product of the Taylor series of the factors we have the following rule: $j^k(fg)(t_0)$ is the product of $f(t_0)+j^k f(t_0)$ and $g(t_0)+j^k g(t_0)$ with the constant term and all terms of degree $>k$ deleted. For example, the 4-jet of $e^t\cos 2t$ at 0 is

$$(1+t+\tfrac{1}{2}t^2+\tfrac{1}{6}t^3+\tfrac{1}{24}t^4)(1-\tfrac{1}{2}4t^2+\tfrac{1}{24}16t^4)$$

with the constant and terms of degree $>4$ deleted, i.e. $t-\frac{3}{2}t^2+\frac{1}{6}t^3-\frac{11}{24}t^4$. Note that we have to include the constant terms in the factors before multiplying them together.

(5) Let $\alpha\colon \mathbb{R},\ 0\to\mathbb{R}^2$ be a unit speed plane curve, where $\alpha(0)=(0,\,0)$, $\alpha'(0)=(1,\,0)$. The $k$-jet of $\alpha$ at 0 is (by definition) the pair of $k$-jets of the components. Exercise 2.28(4) shows that the 4-jet of $\alpha$ at 0 is

$$(t-\tfrac{1}{6}\kappa^2 t^3-\tfrac{1}{8}\kappa\kappa' t^4,\ \tfrac{1}{2}\kappa t^2+\tfrac{1}{6}\kappa' t^3+\tfrac{1}{24}(\kappa''-\kappa^3)t^4),$$

where $\kappa,\ \kappa',\ \kappa''$ are all evaluated at 0.

Similarly 2.36(4) provides a formula for the 4-jet at 0 of a unit speed space curve $\alpha: \mathbb{R}, 0 \to \mathbb{R}^3$ in 'standard position' with $\alpha(0) = (0, 0, 0)$, $T(0) = (1, 0, 0)$, $N(0) = (0, 1, 0)$ (so we assume $\kappa(0) \neq 0$).

Neither of these formulas is affected if we replace $t = 0$ by $t = t_0$ (except that $\kappa$, $\kappa'$ etc. are now evaluated at $t_0$).

(6) We can work out $k$-jets of composites by 'neglecting higher terms'. For example the 4-jet of $\exp(\sin t)$ at 0 is obtained from

$$1 + (t - \tfrac{1}{6}t^3) + \tfrac{1}{2}(t - \tfrac{1}{6}t^3)^2 + \tfrac{1}{6}(t - \tfrac{1}{6}t^3)^3 + \tfrac{1}{24}(t - \tfrac{1}{6}t^3)^4$$

by deleting the constant and terms of degree $> 4$. Hence it is $t + \tfrac{1}{2}t^2 - \tfrac{1}{8}t^4$.

(7) Let $\gamma(t) = (\cos t, b \sin t)$, where $b > 0$, be a parametrization of an ellipse. Let $f: \mathbb{R} \to \mathbb{R}$ be the square of the distance from $(\tfrac{1}{2}, 0)$ to $\gamma(t)$, namely $f(t) = (\tfrac{1}{2} - \cos t)^2 + b^2 \sin^2 t$. For each $t_0$ we consider the 2-jet at $t_0$ of $f$, which is

$$(\sin t_0 + (b^2 - 1)\sin 2t_0)t + (\cos t_0 + 2(b^2 - 1)\cos 2t_0)t^2.$$

Writing this as say $c_1 t + c_2 t^2$, the point $(c_1, c_2) \in \mathbb{R}^2$ will trace out a curve $\delta$ as $t_0$ varies. The curve $\delta$ will cross the $c_2$-axis at points $\delta(t_0)$ where $f$ has a singularity $A_{\geqslant 1}$ at $t_0$, and will pass through the origin where $f$ has an $A_{\geqslant 2}$ at $t_0$. The geometrical interpretation of this is that $f$ has an $A_{\geqslant 1}$ singularity at $t_0$ iff the normal to the ellipse at $\gamma(t_0)$ passes through $(\tfrac{1}{2}, 0)$ and an $A_{\geqslant 2}$ iff $(\tfrac{1}{2}, 0)$ is the centre of curvature of the ellipse at $\gamma(t_0)$. (Compare 2.27.) Figure 3.1 shows the image of $\delta$ for two values of $b$, namely

**Fig. 3.1.** The curve of example (7) for $b = \tfrac{1}{2}$ and $b = \tfrac{1}{2}\sqrt{2}$

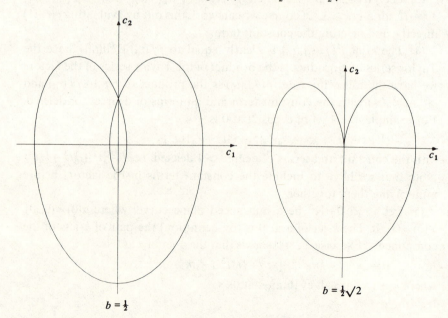

$b = \tfrac{1}{2}$

$b = \tfrac{1}{2}\sqrt{2}$

$b=\frac{1}{2}$ and $b=\frac{1}{2}\sqrt{2}$. In the first case there are four $A_{\geqslant 1}$ singularities (four normals through $(\frac{1}{2}, 0)$) while in the second case two have coincided to give $A_{\geqslant 2}$. (Bearing in mind that the radius of curvature of the ellipse at $(1, 0)$ and $(-1, 0)$ is $b^2$ (see (2.26)(1)), another value of $b$ for which $\delta$ passes through $(0, 0)$ is $\sqrt{(\frac{3}{2})}$. These are the only two values of $b$ for which this happens, as you can check by eliminating $t_0$ between $c_1=0$ and $c_2=0$.)

## 3.13    Exercises

(1) Verify the following 4-jets at 0:

$\exp(1+t^2)$: $et^2+\frac{1}{2}et^4$

$\sin(e^t-1)$: $t+\frac{1}{2}t^2-\frac{5}{24}t^4$.

(What goes wrong with the substitution technique as in (6) above for $\sin(e^t)$?)

$\sin t/1+t^2$: $t-\frac{7}{6}t^3$.

(2) Find the formula for the 4-jet of the unit speed space curve referred to in (5) above.

(3) Repeat (7) above, replacing $(\frac{1}{2}, 0)$ by $(0, 0)$. Sketch the image of the corresponding map $\delta$ and show that, for $b\neq 1$, the image never passes through $(0, 0)$. How many normals pass through $(0, 0)$? (Careful! The image is described *twice* as $t_0$ goes from 0 to $2\pi$.)

(4) Let $F(t, x, y)=\|\gamma(t)-(x, y)\|^2$, where $\gamma$: $\mathbb{R}$, $0\rightarrow\mathbb{R}^2$ is a (regular) plane curve. Let $\gamma(t)=(X(t), Y(t))$. Show that the 1-jet with constant at 0 of $\partial F/\partial x(t, a, b)$ (which will be a function of $t$ only) is

$-2(X(0)-a)-2X'(0)t$.

Show that the 1-jets with constant at 0 of $\partial F/\partial x(t, a, b)$ and $\partial F/\partial y(t, a, b)$ are *independent* polynomials provided $(\gamma(0)-(a, b))\cdot N(0)\neq0$. (What does this condition mean geometrically?)

(5) Let $\gamma$: $\mathbb{R}$, $0\rightarrow\mathbb{R}^2$ be unit speed and define

$F(t, x, y)=(\gamma(t)-(x, y))\cdot T(t)$.

Write $\gamma(t)=(X(t), Y(t))$. Find the 1-jets with constant at 0 of $\partial F/\partial x(t, a, b)$ and $\partial F/\partial y(t, a, b)$ and show that they are independent provided $\kappa(0)\neq0$.

# 4

## *Regular values and smooth manifolds*

'Where there is no imagination
there is no horror.' (*A Study in Scarlet*)

In an attempt to describe the world in which we live it is natural to try to produce mathematical models of the objects about us, so that we can study their geometry, compare their form, possibly predict their growth. Of course we study the static world by modelling it on Euclidean space $E^3$, which we usually identify with $\mathbb{R}^3$ together with its usual distance function. (Indeed we take this model so much for granted that it almost ceases to be a model at all.)

How are we to describe the objects that appear inside this space? One fairly natural method is to model them on solutions of equations $f(x) = c$ for maps $f: \mathbb{R}^3 \to \mathbb{R}^p$ and points $c \in \mathbb{R}^p$. The next natural question to ask is: what type of maps $f$ should we work with? Various suggestions come to mind: polynomial, analytic (given by convergent power series), differentiable, continuous . . . . Each has its own advantages and disadvantages. From our point of view polynomial and analytic functions are too rigid – they hamper the imagination too much. Using merely continuous functions one forgoes the powerful techniques of differential calculus. Differentiable functions are about the right compromise – actually we require our functions to possess derivatives of all orders; such functions are called *smooth* or $C^\infty$. These functions allow sufficient play for the imagination – and (as Conan Doyle knew well, though he stated the converse) that opens the door to some horrors as well. We take a brief look at those in 4.11 below, but rapidly introduce a saviour in the form of a theorem due to A. Sard (4.18). Smooth functions, aided by Sard's theorem, give us a beautiful universe of mathematical objects with which to model the wonders around us.

50

## Smooth maps and parametrized manifolds

### 4.1 Definition

Let $f: \mathbb{R}^n, u \to \mathbb{R}^p$ be a map, so that $f$ is defined on some open set $U$ of $\mathbb{R}^n$ containing the point $u$. Write $f(x) = (f_1(x), \ldots, f_p(x))$, $x = (x_1, \ldots, x_n)$, so that each component $f_i$ is a function $\mathbb{R}^n, u \to \mathbb{R}$. We call $f$ *smooth* provided all partial derivatives of the $f_i$ of all orders exist and are continuous in $U$.

Thus for $f$ smooth, $\partial^2 f_i / \partial x_1 \partial x_2$, $\partial^3 f_i / \partial x_1^3$, etc., all exist and are continuous, and $\partial^2 f_i / \partial x_1 \partial x_2 = \partial^2 f_i / \partial x_2 \partial x_1$, etc.

### 4.2 Definition

Let $f$ as above be smooth and let $v \in U$. The matrix

$$\begin{pmatrix} \dfrac{\partial f_1}{\partial x_1} & \dfrac{\partial f_1}{\partial x_2} & \cdots & \dfrac{\partial f_1}{\partial x_n} \\ \vdots & \vdots & & \vdots \\ \dfrac{\partial f_p}{\partial x_1} & \dfrac{\partial f_p}{\partial x_2} & \cdots & \dfrac{\partial f_p}{\partial x_n} \end{pmatrix}$$

where the partial derivatives are evaluated at $v$, is called the *Jacobian matrix* of $f$ at $v$. The linear map $Df(v): \mathbb{R}^n \to \mathbb{R}^p$ with this matrix is called the *derivative* (in some books the *differential*) of $f$ at $v$.

For example a map $f$, each of whose components $f_i$ is a polynomial function of $x_1, \ldots, x_n$, is certainly smooth. If the components are rational functions then the map is smooth so long as none of the denominators vanishes anywhere in $U$; hence if none vanishes at $u$ then we obtain a smooth map on some open set $U_1 \subset U$ with $u \in U_1$. The functions sin, cos, exp of one variable are smooth and composing these with polynomials $f_i$ (for example $f(x_1, x_2) = (\sin(x_1 x_2), \exp(x_1^2), \cos(x_1 - x_2^4 + 3))$, where $n = 2$ and $p = 3$), gives a smooth map. The composite of two smooth maps, possibly restricted to a smaller open set, is smooth.

### 4.3 Definition

Let $U$ and $V$ be open subsets of $\mathbb{R}^m$. A map $\phi: U \to V$ is called a *diffeomorphism* if (i) $\phi$ is bijective, (ii) both $\phi$ and $\phi^{-1}$ are smooth.

We shall need the following fundamental theorem from advanced calculus – see for example Spivak (1965), theorem 2-11.

### 4.4 Inverse Function Theorem

*Let $\phi: U_1 \to \mathbb{R}^m$ be smooth, with $U_1$ open in $\mathbb{R}^m$ (same $m$!) and $u \in U_1$. Suppose that the Jacobian matrix of $\phi$ at $u$ is a nonsingular matrix. Then* *there exists an open set $U \subset U_1$, with $u \in U$, such that $\phi: U \to \phi(U)$ is a diffeomorphism.* We call $\phi$ a *local diffeomorphism* at $u$.

### 4.5    Examples

(1) Let $m = 1$. Then for $\phi: U \to \mathbb{R}$ ($U$ being open in $\mathbb{R}$, say an open interval containing $u$), the Jacobian matrix is $1 \times 1$ and so nonsingular iff it is nonzero. Thus $\phi$ is a local diffeomorphism iff $\phi'(u) \neq 0$. (Compare 2.16.) For instance, $\phi(x) = x^k$ ($U = \mathbb{R}$, $k$ a positive integer) is a local diffeomorphism at 0 iff $k = 1$, for the Jacobian is $(kx^{k-1})$ for $k > 1$ and just $(1)$ for $k = 1$. But $\phi$ is a local diffeomorphism at $u \neq 0$ for all $k$.

(2) Let $\phi(x, y) = (x, y^2)$ ($U = \mathbb{R}^2$, $m = 2$). The Jacobian matrix at $(x, y)$ is $\begin{pmatrix} 1 & 0 \\ 0 & 2y \end{pmatrix}$ which is nonsingular iff $y \neq 0$. Thus $\phi$ is a local diffeomorphism at any point $(x, y)$ with $y \neq 0$. This map is called a fold: the $(x, y)$-plane is folded over and creased along the axis $y = 0$.

### 4.6    Exercises

(1) Let $\gamma(t) = (t, Y(t))$ be the curve whose image is the graph of the smooth function $y = Y(x)$. Why is $\gamma$ automatically regular for all $t$? Let $\phi(x, y) = (x, y + Y(x))$. Show that $\phi$ is a diffeomorphism $\mathbb{R}^2 \to \mathbb{R}^2$ which takes the $x$-axis onto the image of the curve $\gamma$. Write down $\phi^{-1}$ explicitly. Thus by a global diffeomorphism (defined on the whole of $\mathbb{R}^2$) the graph of any smooth function can be taken to the $x$-axis. Does a similar result hold for curves in $\mathbb{R}^3$?

(2) Let $\gamma: I \to \mathbb{R}^2$ be any (regular) plane curve, with $\gamma(t) = (X(t), Y(t))$, and let $\phi(x, y) = (X(x), y + Y(x))$. Show that $\phi$ is injective iff $X$ is injective. With $\gamma(t) = (t^3, t)$, show that $\phi$ is bijective, but $\phi^{-1}$ is not smooth at points where $x = 0$. Show that for any $\gamma$, $\phi$ is a local diffeomorphism at $(x_0, y_0)$ provided $X'(x_0) \neq 0$, i.e. provided the tangent to the curve at $t = x_0$ is not *vertical*. If $X'(x_0) = 0$, then $Y'(x_0) \neq 0$ (why?); in that case redefine $\phi$ to be $\phi(x, y) = (y + X(x), Y(x))$: this is a local diffeomorphism at $(x_0, y_0)$.

(3) Carry out a similar investigation to that in (2) for a general space curve $\gamma(t) = (X(t), Y(t), Z(t))$, where say $X'(x_0) \neq 0$.

The result of exercise (2) above can be put in the following way. For a (regular) plane curve $\gamma(t) = (X(t), Y(t))$, suppose that $X'(x_0) \neq 0$. Then there is an open set $U$ in $\mathbb{R}^2$ – say an open disc – containing $(x_0, 0)$ which is taken into $\mathbb{R}^2$ by a local diffeomorphism $\phi$ ($\phi(x, y) = (X(x), y + Y(x))$) in such a way that $(x\text{-axis}) \cap U$ is taken to a piece of $\gamma(I)$ corresponding to parameter values $t$ near $x_0$. See fig. 4.1. Notice that $\phi(U)$ may contain other pieces of $\gamma(I)$, with $t$ not close to $x_0$, as in the lower diagram.

In some ways it is more convenient if, in the above situation, $U$ can

**Fig. 4.1.**

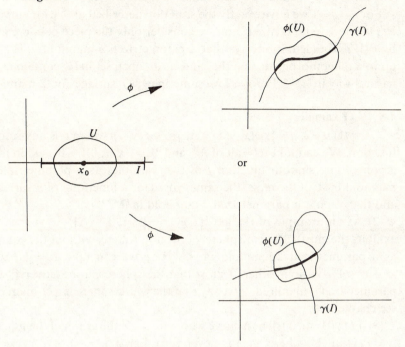

always be chosen so that the image $\phi(U)$ contains just *one* piece of the curve. When this happens we say that the curve is a '1-manifold', and these objects (or their higher dimensional cousins) will occupy us in chapter 8. For the moment we shall consider a slightly more restricted situation, where we have just one set $U$ in mind.

In $\mathbb{R}^{n+q} = \mathbb{R}^n \times \mathbb{R}^q$ let $U$ be an open set containing $w = (u_1, \ldots, u_n, 0, \ldots, 0)$, intersecting the set $\mathbb{R}^n \times \{0\}$, which we identify with $\mathbb{R}^n$, in say $W$. We shall take this $W$ to be connected, or consider one connected component of $W$. Let $\phi: U \to \mathbb{R}^{n+q}$ be smooth, with $\phi: U \to \phi(U)$ a diffeomorphism.

### 4.7 Definition

$M = \phi(W)$ is called a *parametrized n-manifold* in $\mathbb{R}^{n+q}$ and the restriction $\gamma = \phi|W$ is called a *parametrization* of $M$. The integer $q$ is called the *codimension* of $M$ in $\mathbb{R}^{n+q}$.

Since $\gamma(x_1, \ldots, x_n) = \phi(x_1, \ldots, x_n, 0, \ldots, 0)$, we have $\dfrac{\partial \gamma_i}{\partial x_j}(w) = \dfrac{\partial \phi_i}{\partial x_j}(w)$ when $1 \leqslant j \leqslant n$ (and $1 \leqslant i \leqslant n+q$). Hence the Jacobian matrix of $\gamma$ at $w$ consists of the first $n$ columns of the Jacobian matrix of $\phi$ at $w$, and consequently the former has rank $n$. Such a $\gamma$ is called an 'embedding'.

The set $M$ is also called an 'embedded submanifold' of $\mathbb{R}^{n+q}$ (fig. 4.2).

For $n=q=1$ we have exactly the situation described above for curves in $\mathbb{R}^2$, where $W=(x\text{-axis})\cap U$ and we consider only the piece of curve given by $\phi(W)$. The definition says that a parametrized $n$-manifold in $\mathbb{R}^{n+q}$ is, up to a diffeomorphism, just the same as an open set in $\mathbb{R}^n$ contained in a natural way in $\mathbb{R}^{n+q}$. For $n=2$ we sometimes say 'surface' for '2-manifold'.

## 4.8    Examples

(1) $n=2$, $q=1$, $\phi(x, y, z)=(x, y, z+f(x, y))$ where $f$ is any smooth function. We can take $U=$ all of $\mathbb{R}^3$, and $W=\{(x, y, 0)\}$. This exhibits the graph of any smooth function $f$ of two variables as a parametrized 2-manifold in $\mathbb{R}^3$. Of course, the same works for a function $f$ of $n$ variables, and the graph is a parametrized $n$-manifold in $\mathbb{R}^{n+1}$.

(2) As an example of (1), let $\phi(x, y, z)=(x, y, z+\sqrt{(1-x^2-y^2)})$. This exhibits the 'northern hemisphere' of the unit sphere, without the equator, as a parametrized 2-manifold in $\mathbb{R}^3$. Here we can take $U=\{(x, y, z):$ $x^2+y^2<1\}$. Various other $\phi$ show that the sphere can be 'covered' with parametrized 2-manifolds – we say that the whole sphere is a '2-manifold'; see chapter 8.

(3) Let $\gamma(t)=(t, Y(t))$ be a curve whose image is the graph of the function $y=Y(x)$ in $\mathbb{R}^2$, where $Y(x)$ is never zero. Let $\phi(x, y, z)=(x, Y(x)\cos y, z+Y(x)\sin y)$, where we take $U=\{(x, y, z): 0<y<\pi\}$. This exhibits part of the surface of revolution generated by rotating the graph about the $x$-axis as a parametrized 2-manifold in $\mathbb{R}^3$. (You should check that the conditions are satisfied.) Other parts of the surface can be parametrized by shifting $z$ to the second component of $\phi$. (What happens if $Y(x)$ is sometimes zero? Try $Y(x)=x$ and draw a diagram.)

Another interesting example is obtained by taking any smooth $\gamma: W \to \mathbb{R}^3$ where $W$ is open and connected in $\mathbb{R}^2$, which is an *immersion*, i.e. the Jacobian matrix has maximal rank, namely 2, at each $w \in W$. (Equivalently,

**Fig. 4.2.** Parametrized manifold

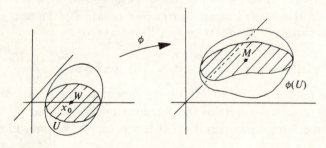

$D\gamma(w)$ is injective for each $w$.) Given such a $\gamma$, and $w \in W$, two rows of the $3 \times 2$ Jacobian matrix must be independent; let us assume it is the first two rows. (Similar arguments will hold for other pairs of independent rows.) Then $\phi(x, y, z) = (\gamma_1(x, y), \gamma_2(x, y), z + \gamma_3(x, y))$ has Jacobian matrix at $(x_0, y_0, 0)$, where $w = (x_0, y_0)$

$$\begin{pmatrix} \text{Jacobian} & 0 \\ \text{of } \gamma & 0 \\ \text{at } w & 1 \end{pmatrix}$$

which will be nonsingular. Hence $\phi$ is *locally* a diffeomorphism, by the inverse function theorem. That is, for some open set $U \subset \mathbb{R}^3$, containing $(x_0, y_0, 0)$, we have $\phi | U$ a diffeomorphism onto its image. Thus for some open connected $W_1 \subset W$, containing $(x_0, y_0)$, we have $\gamma(W_1) = \phi(W_1)$ is a parametrized 2-manifold in $\mathbb{R}^3$. We summarize this as follows.

### 4.9 Proposition

*The image of an immersion* $\gamma \colon W \to \mathbb{R}^3$ *(W open and connected in* $\mathbb{R}^2$*) is locally a parametrized 2-manifold in* $\mathbb{R}^3$*, in the sense that for each* $w \in W$ *there exists a connected open set* $W_1 \subset W$*, with* $w \in W_1$*, such that* $\gamma(W_1)$ *is a parametrized 2-manifold.* □

### 4.10 Exercises

(1) Let $W = \{(x, y) \in \mathbb{R}^2 : 0 < y < \pi\}$ and define $\gamma(x, y) = (r \cos x \sin y, r \sin x \sin y, r \cos y)$ for a fixed $r > 0$. Show that $\gamma$ is an immersion. The image $\gamma(W)$ is a sphere in $\mathbb{R}^3$ minus the north and south poles.

(2) Given a regular curve $\alpha(t) = (X(t), Y(t))$ find conditions for $\gamma(x, y) = (X(x), Y(x)\cos y, Y(x)\sin y)$ to be an immersion (here $\gamma$ is defined on $I \times \mathbb{R}$ where $I$ is the domain of $\alpha$. Taking $\alpha(t) = (\cos t, \sin t + 2)$ identify the image of $\gamma$ as a torus of revolution in $\mathbb{R}^3$. (Compare 4.8(3).)

(3) Let $\alpha \colon I \to \mathbb{R}^3$ be a unit speed space curve. Define $\gamma \colon I \times \mathbb{R}_+ \to \mathbb{R}^3$ by $\gamma(x, y) = \alpha(x) + y\alpha'(x)$ (here $\mathbb{R}_+ = \{y \in \mathbb{R} : y > 0\}$). Show that $\gamma$ is an immersion provided the curvature $\kappa(t)$ is never zero on $I$. (The same applies to $y < 0$, and the whole image in $\mathbb{R}^3$ is the union of all tangent lines to $\alpha$, except points of contact, called the *tangent developable* of $\alpha$. See fig. 7.5.)

(4) Generalize 4.9 to immersions $\gamma \colon U \to \mathbb{R}^{n+q}$ where $U$ is open in $\mathbb{R}^n$ (thus the Jacobian of $\gamma$ is assumed to have rank $n$ at each point $u \in U$). As an example, consider $n = q = 2$, $\gamma(x, y) = (\cos x, \sin x, \cos y, \sin y)$, whose image is called a 'flat torus' in $\mathbb{R}^4$.

## Regular values

There is another good way of specifying (say) a curve in $\mathbb{R}^2$ or a surface in $\mathbb{R}^3$ besides by parametrization. Curves are often specified by equations $f(x, y) = 0$; similarly a surface could be given by $f(x, y, z) = 0$. A curve in $\mathbb{R}^3$ might be specified by *two* equations $f(x, y, z) = g(x, y, z) = 0$. What do such solution sets look like? The answer, even for the solution set of one equation $f(x_1, \ldots, x_n) = 0$, is extremely discouraging.

### 4.11    Theorem (H. Whitney)

*Any closed set in $\mathbb{R}^n$ occurs as the solution set $f^{-1}(0)$ $= \{x \in \mathbb{R}^n : f(x) = 0\}$ for some smooth function $f : \mathbb{R}^n \to \mathbb{R}$.* (See Bröcker and Lander (1975).)

Closed sets can be quite awful (think of the Cantor set, if you can bear to) so along with curves and other pleasant creatures we admit a host of monsters when we look at solution sets $f(x) = 0$. (Compare the introductory remarks to this chapter.) But a remarkable theorem (definitely a three-pipe theorem) comes to the rescue here. We stick to a fixed smooth function $f$ but look at solution sets $f(x) = c$ for constants $c$; the theorem, due to A. Sard, will tell us that even though we cannot hope to describe the local structure of $f^{-1}(c)$ for arbitrary $c$, for *most* values of $c$ the local structure is very simple – in fact it is that of a parametrized manifold. First a definition.

### 4.12    Definition

Let $f : \mathbb{R}^m \to \mathbb{R}^p$ be smooth (the arrow $\to$ indicates that the domain of $f$ is an open subset of $\mathbb{R}^m$). A point $x$ of $\mathbb{R}^m$ is called a *regular point* of $f$, and $f$ is called a *submersion at $x$*, provided $Df(x)$ is surjective. This is the same as saying that the Jacobian matrix of $f$ at $x$ has rank $p$ (which is only possible if $p \leqslant m$). A *regular value* of $f$ is a point $c \in \mathbb{R}^p$ such that every $x$ in the domain of $f$ with $f(x) = c$ is a regular point.

The term *critical point* is used for any point $x \in \mathbb{R}^m$ for which the rank of the linear map $Df(x)$ ($=$ rank of the Jacobian matrix) falls below its largest possible value, namely $\min(m, p)$. Likewise a *critical value* is any $f(x) \in \mathbb{R}^p$ where $x$ is a critical point. For $m \geqslant p$ 'critical' and 'non-regular' are the same, since then $\min(m, p) = p$.

Note that for $p = 1$ the condition for $x$ to be a non-regular point is $\partial f / \partial x_1 = \cdots = \partial f / \partial x_m = 0$ at $x$.

Notice also one peculiarity of the definition: if there does not exist $x \in \mathbb{R}^m$ with $f(x) = c$, then $c$ is automatically a regular value of $f$ (for there is no $x$ where we need to check that $Df(x)$ is surjective). It could be argued that such $c$ are not values of $f$ at all, so how could they be 'regular values'? Alas, even in mathematics the terminology is sometimes illogical!

**4.13    Examples**

(1) Let $m=p=1, f(x)=x^2$. Then any $c \neq 0$ is a regular value (for $c<0$ compare the peculiarity noted above). For $Df(x)$ has matrix $(2x)$ which has rank 1 (i.e. is nonzero) iff $x \neq 0$: the only non-regular point of $f$ is 0.

(2) Let $f: \mathbb{R}^2 \rightarrow \mathbb{R}$ be $f(x, y)=2x^2+3y^2$. Then the Jacobian matrix is $(4x \quad 6y)$ which has rank 1 unless $x=y=0$. So any $c \neq 0$ is a regular value. For $c>0$, $f^{-1}(c)$ is an ellipse in the plane.

(3) Let $f: \mathbb{R}^2 \rightarrow \mathbb{R}$ be $f(x, y)=x^3+y^3+xy$. Then $\partial f/\partial x=\partial f/\partial y=0$ gives $x=y=0$ or $x=y=-\frac{1}{3}$, so that $(0, 0)$ and $(-\frac{1}{3}, -\frac{1}{3})$ are the only non-regular (critical) *points* of $f$. Since $f(0, 0)=0$, $f(-\frac{1}{3}, -\frac{1}{3})=1/27$ it follows that any $c$ other than 0 or 1/27 is a regular *value* of $f$. Also 0 is a regular value of $f|(\mathbb{R}^2-\{(0, 0)\})$ and 1/27 is a regular value of $f|(\mathbb{R}^2-\{(-\frac{1}{3}, -\frac{1}{3})\})$. Figure 4.3 shows $f^{-1}(c)$ for some values of $c$.

(4) Let $f: \mathbb{R}^3 \rightarrow \mathbb{R}^2$ be $f(x, y, z)=(xy, xz)$. The Jacobian of $f$ is $\begin{pmatrix} y & x & 0 \\ z & 0 & x \end{pmatrix}$ which has rank 2 unless *all* $2 \times 2$ minors are zero, i.e. unless $xz=xy=x^2=0$, which is equivalent to $x=0$. Since $f(0, y, z)=(0, 0)$, any point $(u, v) \in \mathbb{R}^2$ other than $(0, 0)$ is a regular value.

We shall now proceed in two directions: first, to examine $f^{-1}(c)$ for a regular value $c$; and second, to see whether regular values are common or uncommon. For the first we shall use the following consequence of the Inverse Function Theorem.

**4.14    Implicit Function Theorem**

*Let* $f: \mathbb{R}^{n+q}, (a, b) \rightarrow \mathbb{R}^q$ *be a smooth map, defined on a neighbourhood of* $(a, b) \in \mathbb{R}^n \times \mathbb{R}^q = \mathbb{R}^{n+q}$, *with* $f(a, b)=c$. *Write* $(x_1, \ldots, x_n, y_1, \ldots, y_q)$

**Fig. 4.3.** Level curves $x^3+y^3+xy=c$ (broken lines are axes)

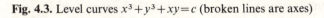

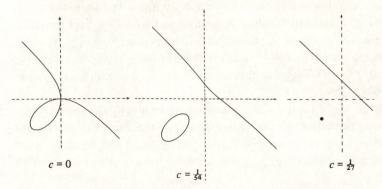

$c = 0$          $c = \frac{1}{54}$          $c = \frac{1}{27}$

*for coordinates in* $\mathbb{R}^{n+q}$ *and consider the* $q \times q$ *matrix*

$$\begin{pmatrix} \dfrac{\partial f_1}{\partial y_1} & \cdots & \dfrac{\partial f_1}{\partial y_q} \\ \vdots & & \vdots \\ \dfrac{\partial f_q}{\partial y_1} & \cdots & \dfrac{\partial f_q}{\partial y_q} \end{pmatrix}$$

*If this matrix is nonsingular at* $(a, b)$ *then there exist neighbourhoods A of a in* $\mathbb{R}^n$, *B of b in* $\mathbb{R}^q$ *such that for all x in A there is a unique point* $g(x)$ *in B with* $f(x, g(x)) = c$. *Furthermore the map* $x \mapsto g(x)$ *is smooth.*

**Proof** Let $F: \mathbb{R}^n \times \mathbb{R}^q, (a, b) \to \mathbb{R}^n \times \mathbb{R}^q$ be defined by $F(x, y) = (x, f(x, y))$. It is easy to check, using the assumed nonsingularity of the above $q \times q$ matrix at $(a, b)$, that the Jacobian matrix of $F$ is nonsingular at $(a, b)$. So, by the Inverse Function Theorem, $F$ has a local inverse $H$, defined on a neighbourhood of $(a, c)$. Using the form of $F$, $H$ must have the form $H(x, y) = (x, h(x, y))$ for some smooth $h$. Let us choose a neighbourhood of $(a, b)$ of the form $A \times B$ such that $F|A \times B$ has a smooth inverse. Then, for $(x, y) \in A \times B$, $f(x, y) = c \Leftrightarrow F(x, y) = (x, c) \Leftrightarrow (x, y) = H(x, c) = (x, h(x, c))$. Thus the unique $y$ is $h(x, c)$, and $g(x) = h(x, c)$, which is smooth. $\qquad\square$

The theorem says that, provided the matrix $(\partial f_i/\partial y_j)$ is nonsingular at $(a, b)$, we can solve the equations $f(x, y) = c$ (where $c = f(a, b)$) locally for $y_1, \ldots, y_q$ as smooth functions of $x_1, \ldots, x_n$. Another way of thinking of this is to say that the set $f^{-1}(c)$ can be parametrized, near $(a, b)$, just using $x_1, \ldots, x_n$ as parameters. As always, this is easiest to see when $n$ and $q$ are small, as in the following examples.

**4.15        Examples**
        (1) $n = q = 1$, $f(x, y) = x^2 + y^2$. Provided $c > 0$, we can solve for $y$ as a smooth function of $x$ close to $(a, b) \in \mathbb{R}^2$ on the circle $f^{-1}(c)$ when the $1 \times 1$ matrix $(2b)$ is nonsingular, i.e. when $b \neq 0$ (fig. 4.4). Likewise we can solve for $x$ as a smooth function of $y$ close to $(a, b)$ on $f^{-1}(c)$ provided $a \neq 0$. (Here the functions can be written down explicitly: $y = \pm \sqrt{(c - x^2)}$, the sign being that of $b$; $x = \pm \sqrt{(c - y^2)}$, the sign being that of $a$. Of course when $c < 0$, $f^{-1}(c)$ is *empty* so there is no point $(a, b)$ at which to say anything!)

        (2) For the $f$ of 4.13(4) we can solve for $y$ and $z$ locally as functions of $x$ on $f^{-1}(u, v)$ provided the last two columns of the Jacobian matrix are independent, i.e. provided $x \neq 0$. Taking a fixed $(u, v) \neq (0, 0)$, *no* point of $f^{-1}(u, v)$ has the form $(0, y, z)$, so we can solve for $y, z$ in terms of $x$ at *all*

**Fig. 4.4.** Solving for $y$ in terms of $x$ on the circle

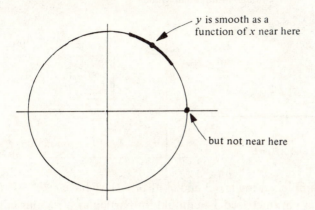

$y$ is smooth as a function of $x$ near here

but not near here

points of $f^{-1}(u, v)$. This is in any case obvious, since $y = u/x$, $z = v/x$. Now put $(u, v) = (0, 0)$. We can still solve for $y$ and $z$ on $f^{-1}(0, 0)$ so long as $x \neq 0$: in fact we just get $y = z = 0$. Of course $f^{-1}(0, 0)$ consists of this axis together with the plane $x = 0$, and we cannot possibly express $y$ and $z$ as functions of $x$ on the plane: they are arbitrary there, while $x$ is fixed.

Now let $f: \mathbb{R}^m, v \to \mathbb{R}^q$, $c$ be a submersion at $v$, i.e. $Df(v)$ surjective. We show that, near $v$, $f^{-1}(c)$ is a parametrized $m - q$ manifold in $\mathbb{R}^m$. In fact, since the Jacobian matrix of $f$ at $v$ has rank $q$, select $q$ columns of this matrix which are independent. For simplicity take these to be the last $q$ columns (a permutation of coordinates in $\mathbb{R}^m$ will achieve this), so that splitting $\mathbb{R}^m$ as $\mathbb{R}^n \times \mathbb{R}^q$ ($n = m - q \geqslant 0$) and writing $v = (a, b) \in \mathbb{R}^n \times \mathbb{R}^q$ the hypotheses of the implicit function theorem are satisfied at $(a, b)$. Let $g: A \to B$ be as in the theorem, so that $f(x, g(x)) = c$ for all $x \in A$. Define $\psi: A \times B \to \mathbb{R}^m$ by

$$\psi(x, y) = (x, y_1 - g_1(x), \ldots, y_q - g_q(x)) = (x, y - g(x)).$$

Certainly $\psi$ is a diffeomorphism onto its image $U$ (the inverse $\phi = \psi^{-1}$ is $\phi(x, y) = (x, y + g(x))$). Also $(x, y) \in f^{-1}(c)$ just when $g(x) = y$ so $\psi$ takes the part of $f^{-1}(c)$ lying in $A \times B$ to the intersection of $\mathbb{R}^n \times \{0\}$ with $U$ (fig. 4.5). Hence we have the following.

### 4.16 Proposition

*Let $f: \mathbb{R}^m, v \to \mathbb{R}^q$, $c$ be a submersion at $v$. Then there is a neighbourhood $V$ of $v$ in $\mathbb{R}^m$ such that $f^{-1}(c) \cap V$ is a parametrized $(m-q)$-manifold in $\mathbb{R}^m$.*

*If $c$ is a regular value of $f: \mathbb{R}^m \twoheadrightarrow \mathbb{R}^q$ then for every $v \in f^{-1}(c)$ the above conclusion holds.* $\qquad\square$

($f^{-1}(c)$ is then a 'smooth $m - q$ manifold'; more of that in chapter 8.)

**Fig. 4.5.** Inverse image of a regular value

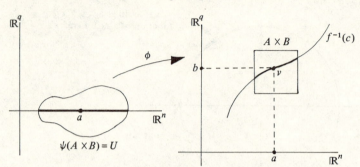

$\psi(A \times B) = U$

Thus, for example, taking $m=2$ and $q=1$ $(n=1)$ we have $f: \mathbb{R}^2 \rightarrowtail \mathbb{R}$ and $f^{-1}(c)$ will be a parametrized 1-manifold (in particular a regular curve) near $(a, b) \in f^{-1}(c)$ provided $\partial f/\partial x$ or $\partial f/\partial y$ is nonzero at $(a, b)$. If $\partial f/\partial x \neq 0$ then we can use $y$ as a local parameter on $f^{-1}(c)$ and if $\partial f/\partial y \neq 0$ then we can use $x$.

## 4.17    Exercises

(1) Let $f: \mathbb{R}^2 \to \mathbb{R}$ be $f(x, y) = 2y^2 - x^5 - x^4 y$. Show that 0 is not a regular value of $f$, but that it is a regular value of $f|\mathbb{R}^2 - \{(0, 0)\}$. Deduce that $V = f^{-1}(0) - \{(0, 0)\}$ is a parametrized 1-manifold in a neighbourhood of any of its points (a 'smooth 1-manifold'). In a neighbourhood of $(1,1) \in V$ show that $V$ can be parametrized by $x$ and calculate $y'$, $y''$ and the curvature of $V$ (using the parametrization by $x$) at $(1, 1)$. (Since $y$ is a function of $x$, you can simply differentiate $2y^2 - x^5 - x^4 y = 0$ with respect to $x$ in order to find $y'$, $y''$.)

(2) Let $f: \mathbb{R}^2 \to \mathbb{R}$ be $f(x, y) = x^2 - y^3$. Show that 0 is not a regular value of $f$ but that 0 is a regular value of $f|\mathbb{R}^2 - \{(0, 0)\}$. Deduce that $f^{-1}(0) - \{(0, 0)\}$ is a parametrized 1-manifold (in particular a regular curve) in a neighbourhood of any of its points. Harder: show that there is no local diffeomorphism $\phi: \mathbb{R}^2, (0, 0) \to \mathbb{R}^2, (0, 0)$ taking the $x$-axis near $(0, 0)$ to the set $f^{-1}(0)$ near $(0, 0)$.

(3) Let $f: \mathbb{R}^3 \to \mathbb{R}$ be $f(x, y, z) = x^2 z - y^2$. Show that $f_1 = f|\mathbb{R}^3 - (z$-axis) has 0 as a regular value. Which two of $x$, $y$, $z$, are local coordinates on $f_1^{-1}(0)$ near $(1, 0, 0)$? Let $g: \mathbb{R}^2 \to \mathbb{R}^3$ be $g(u, v) = (u, uv, v^2)$. Where is $g$ an immersion? (Compare 4.9.) How does the image of $g$ compare with $f_1^{-1}(0)$? The 'surface' $f^{-1}(0)$ is called Whitney's umbrella and is illustrated in fig. 4.6.

(4) Let $f: \mathbb{R}^3 \to \mathbb{R}^2$ be $f(x, y, z) = (xy - z^2, xz - y^2)$. Show that the Jacobian of $f$ has rank $<2$ at $(x, y, z)$ iff $(x, y, z) = (-2\lambda, \lambda, \lambda)$ for some $\lambda \in \mathbb{R}$. Write down $f(-2\lambda, \lambda, \lambda)$ and deduce that $(u, v) \in \mathbb{R}^2$ is a regular value of $f$ unless $u = v \leqslant 0$. Show that $(0, 0)$ is a regular value of $f|\mathbb{R}^3 - \{(0, 0, 0)\}$; also show directly from $f$ that $f^{-1}(0, 0)$ consists of two straight lines through the

**Fig. 4.6.** Whitney's umbrella

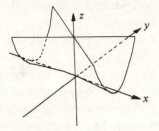

origin. Show that each of $f^{-1}(1, 0)$ and $f^{-1}(1, 1)$ is, in a neighbourhood of each of its points, a parametrized 1-manifold in $\mathbb{R}^3$. Show that, on $f^{-1}(1, 0)$, $z$ can always be used as a local parameter (i.e. $x$ and $y$ can be locally expressed in terms of $z$). Can $x$ always be used as a local parameter?

(5) Let $W \subset \mathbb{R}^2$ be an interval of the $x$-axis containing $(a, 0)$ and let $U$ be an open set in $\mathbb{R}^2$ meeting the $x$-axis in $W$. Let $\phi: U \to \mathbb{R}^2$ be smooth, with $\phi$ a diffeomorphism onto its image $V$. Thus $\phi(W)$ is a parametrized 1-manifold in $\mathbb{R}^2$. Define $f: V \to \mathbb{R}$ by $f(x, y) = \pi(\phi^{-1}(x, y))$ where $\pi$ is 'projection on the second factor', i.e. $\pi(u, v) = v$ for all $(u, v) \in \mathbb{R}^2$. Show that 0 is a regular value of $f$ and that $f^{-1}(0)$ coincides with $\phi(W)$. Thus *any parametrized 1-manifold in $\mathbb{R}^2$ is locally the inverse image of a regular value.*

(6) Generalize (5) to show that *any parametrized n-manifold in $\mathbb{R}^{n+q}$ is locally locally the inverse image of a regular value of a map $f: \mathbb{R}^{n+q} \to \mathbb{R}^q$.*

(7) Real $2 \times 2$ matrices $\begin{pmatrix} p & q \\ r & s \end{pmatrix}$ can be identified with $\mathbb{R}^4$ (coordinates $(p, q, r, s)$). Show that 0 is a regular value of the map $\mathbb{R}^4 - \{(0, 0, 0, 0)\} \to \mathbb{R}$ given by the determinant $ps - qr$. What does this say about the set of matrices of rank 1 as a subset of $\mathbb{R}^4$? What about $3 \times 3$ matrices of rank 2?

(8) Real $2 \times 3$ matrices $\begin{pmatrix} p & q & r \\ s & t & u \end{pmatrix}$ can be identified with $\mathbb{R}^6$. Let $X \subset \mathbb{R}^6$ consist of those matrices of rank 1. Do you think that $X$ is, close to any given $A \in X$, a parametrized manifold in $\mathbb{R}^6$? If so, of what dimension? (Do you think that the fact that a matrix has rank 1 enables you to express some of the entries as functions of the others?)

Are regular values the rule or the exception? The answer is provided by the following theorem of A. Sard (see Milnor (1965) and the appendix).

## 4.18 Sard's Theorem

*Let $f: \mathbb{R}^n \to \mathbb{R}^p$ (defined on an open subset of $\mathbb{R}^n$) be a smooth map. Then the set of non-regular values of $f$ is a null set in $\mathbb{R}^p$, i.e. it has measure zero in the sense of Lebesgue measure.*

When $p = 1$ this means that, for every $\varepsilon > 0$, the set $S$ of non-regular values can be covered with a collection of open intervals of total length less than $\varepsilon$. In all the examples so far, $S$ has actually been *finite*, hence certainly of measure zero. It is not hard to construct functions, $\mathbb{R} \to \mathbb{R}$ say, with a countable infinity of non-regular values. (For example, tilting the graph of $y = \sin x$ in the plane through a small angle about $(0, 0)$ gives the graph of such a function.)

When $p = 2$ Sard's Theorem says that for any $\varepsilon > 0$ $S$ can be covered by a collection of open squares of total area $< \varepsilon$. The complement of the null set $S$ is in fact *dense*, that is, every point of $\mathbb{R}^p$ is arbitrarily close to a regular value of $f$. (See the appendix on Sard's Theorem.) So if we happen to be at a nasty non-regular value then there are plenty of friendly regular values close at hand. In this sense 'almost all' values are regular and, for a given $f$, 'almost all' sets $f^{-1}(c)$ will be parametrized manifolds in a neighbourhood of every point.

For the time being we regard Sard's Theorem as an antidote to Whitney's Theorem 4.11, but here is a rapid application.

Let $f : I \to \mathbb{R}$ be smooth and define $\Phi : I \times \mathbb{R}^2 \to \mathbb{R}$ by $\Phi(t, u) = f(t) + u_1 t + u_2 t^2$. Then $\Phi$ is a family of functions $\Phi_u$, where $\Phi_u(t) = \Phi(t, u)$ and $\Phi_0 = f$. Now let $g : I \to \mathbb{R}^2$ be $g(t) = (-f'(t) + tf''(t), -\frac{1}{2}f''(t))$. It is easy to show that $u \in \text{image}(g)$ if and only if there exists $t \in I$ where $\Phi_u'(t) = \Phi_u''(t) = 0$. Since image$(g)$ is the set of critical values of $g$ in this case, Sard's Theorem says that for all $u$ outside a null set in $\mathbb{R}^2$, $\Phi_u$ has *no* singularities other than of type $A_1$. Thus in particular *there are arbitrarily small 'perturbations' of $f$ which have only $A_1$ singularities.*

We can interpret the above statement more vaguely as saying that we do not 'expect' a function of one variable to have worse than $A_1$ singularities. Indeed an $A_{\geqslant 2}$ singularity imposes *two* conditions on *one* variable $t$, and it is 'unlikely' that these can be simultaneously satisfied for any $t$.

More generally $q$ equations $f_1(x) = \cdots = f_q(x) = 0$ in $p$ variables $x = (x_1, \ldots, x_p)$ give a map $f : \mathbb{R}^p \to \mathbb{R}^q$ and almost all points $c \in \mathbb{R}^q$ are regular values of $f$, by Sard's Theorem. Thus we 'expect' 0 to be a regular value and, if $q > p$, this implies that $f^{-1}(0)$ is *empty*. Thus we 'expect' $q$ equations in $p < q$ unknowns to have no solutions, using Sard's Theorem. More precise versions of this line of reasoning come from 8.17 (Thom's Transversality Lemma) onwards.

## Tangent spaces

In chapter 2 we made a great deal of use of the tangent vector to a curve. We now introduce the corresponding notion for a parametrized $n$-manifold in $\mathbb{R}^{n+q}$. A good example to keep in mind all the time is $n = 2$,

$q = 1$: a surface in $\mathbb{R}^3$, where at every point there is a tangent *plane*. This is in some sense the best 'linear approximation' to the surface at that point.

Let $v \in \mathbb{R}^n$. A *vector at v* is a pair $(v, x)$ where $x \in \mathbb{R}^n$. We usually write $x_v$ or $(x_1, \ldots, x_n)_v$ for $(v, x)$ and write just $x$ for $x_0$, where 0 is the origin of $\mathbb{R}^n$. Geometrically $x_v$ is thought of as an arrow *starting* at $v$ and *ending* at $x + v$ (fig. 4.7). Thus it does matter here where a vector starts from – i.e. where a vector is *based*. The collection of all vectors based at $v$, i.e. $\{x_v : x \in \mathbb{R}^n\}$, is denoted by $\mathbb{R}^n_v$ and is called the *tangent space to $\mathbb{R}^n$ at v*. Also $x_v$ is called a *tangent vector to $\mathbb{R}^n$ at v*.

Clearly $\mathbb{R}^n_v$ is a vector space isomorphic to $\mathbb{R}^n$, with vector space structure defined by $x_v + y_v = (x + y)_v$, $\lambda(x_v) = (\lambda x)_v$ for $\lambda \in \mathbb{R}$. (We do not add $x_v$ to $y_u$ for $u \neq v$.) Think of $\mathbb{R}^n_v$ as another copy of $\mathbb{R}^n$ sitting with its origin at the point $v$ of the first $\mathbb{R}^n$ (fig. 4.8). The endpoints of the various vectors $x_v$ are at $x + v \in \mathbb{R}^n$ and subspaces of $\mathbb{R}^n_v$ are just subspaces of $\mathbb{R}^n$ translated by the vector $v$.

**Fig. 4.7.** A vector at $v$

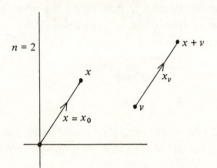

**Fig. 4.8.** Tangent space to $\mathbb{R}^2$ at $v$

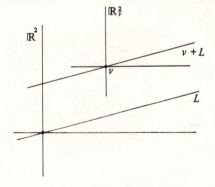

**4.19     Example**

Let $\gamma: I \to \mathbb{R}^n$ be a (regular) curve, and let $t \in I$, $\gamma(t) = v$. Then $(\gamma'(t))_v$ is a vector at $v$, and it points along the tangent line to the curve for the parameter value $t$ (fig. 4.9). (In other words, the tangent line is the image of the linear map $D\gamma(t)$, translated so as to pass through $v$.)

Consider the map $T\gamma(t): \mathbb{R}_t \to \mathbb{R}^n_v$ defined by $s_t \mapsto (s\gamma'(t))_v$, where $s \in \mathbb{R}$. Note that $\gamma'(t) = D\gamma(t)(1)$, where $D\gamma(t): \mathbb{R} \to \mathbb{R}^n$ is the derivative of $\gamma$ at $t$, i.e. the linear map with matrix the column vector $\mathrm{col}(\gamma'_1(t), \ldots, \gamma'_n(t))$. The linearity of $D\gamma(t)$ implies that $s\gamma'(t) = D\gamma(t)(s)$. The map $T\gamma(t)$ is linear, and its image consists of all the vectors at $v$ pointing along the tangent line to $\gamma$. Thus the image of $T\gamma(t)$ can be regarded as the tangent line to the curve.

Of course if $\gamma(t_1) = \gamma(t_2) = v$, so that $\gamma(I)$ has a crossing point, then there will be two tangent lines at $v$, one given by parameter value $t_1$ and the other by parameter value $t_2$. For parametrized 1-manifolds this problem does not arise.

**4.20     Definition**

Let $f: \mathbb{R}^n, u \to \mathbb{R}^p, v$ be smooth (thus $f$ is defined on an open set containing $u$ and $f(u) = v$). The *tangent map* of $f$ at $u$ is the linear map

$$Tf(u): \mathbb{R}^n_u \to \mathbb{R}^p_v$$

whose matrix is the Jacobian matrix of $f$ at $u$. Thus

$$Tf(u)(x_u) = (Df(u)(x))_v \quad \text{for all } x_u \in \mathbb{R}^n_u.$$

Now let $\phi: U \to \mathbb{R}^{n+q}$ be as in 4.7, with $M = \phi(W)$ a parametrized $n$-manifold in $\mathbb{R}^{n+q}$, $W$ being $U \cap \mathbb{R}^n \times \{0\}$ so that $W$ is regarded as a subset of $\mathbb{R}^n$. Let $\gamma = \phi | W$ and $w \in W$.

**4.21     Definition**

The *tangent space* $M_v$ to $M$ at $v = \gamma(w)$ is the image of the tangent map

$$T\gamma(w): \mathbb{R}^n_w \to \mathbb{R}^{n+q}_v.$$

**Fig. 4.9.** Tangent vector to a curve

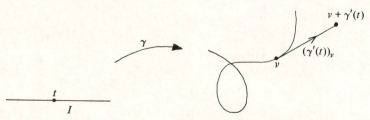

In this chapter we shall consider only parametrized manifolds given by explicit parametrizations $\gamma$; thus we are not concerned with different parametrizations of the same set $M$ in $\mathbb{R}^{n+q}$. See chapter 8 for the latter problem.

Let $\tau_j$ denote the vector $(\partial \gamma_1/\partial x_j, \ldots, \partial \gamma_{n+q}/\partial x_j)$, evaluated at $w$, which forms the $j$th column of the Jacobian matrix of $\gamma$ at $w$. Then $\tau_j \in \mathbb{R}^{n+q}$ and $\tau_1, \ldots, \tau_n$ span the image of the derivative $D\gamma(w)$: $\mathbb{R}^n \to \mathbb{R}^{n+q}$. Likewise $(\tau_1)_v, \ldots, (\tau_n)_v$ span the image of the tangent map $T\gamma(w)$: $\mathbb{R}^n_w \to \mathbb{R}^{n+q}_v$, i.e. they span the tangent space $M_v$. The *tangent space can be thought of as the image of the derivative, translated out to the point $v$* (fig. 4.10).

## 4.22    Examples

(1) Let $f: \mathbb{R}^2 \to \mathbb{R}$ be smooth, and take $\phi: \mathbb{R}^3 \to \mathbb{R}^3$ to be $\phi(x, y, z) = (x, y, z + f(x, y))$, so that $\gamma(x, y) = (x, y, f(x, y))$ has image the graph of the function $f$. (Compare 4.8(1).) The tangent plane $M_v$ to $M = \gamma(\text{domain } f)$ at $v = \gamma(a, b)$ is spanned by $(\tau_1)_v, (\tau_2)_v$ where

$$\tau_1 = (1, 0, \partial f/\partial x), \tau_2 = (0, 1, \partial f/\partial y)$$

evaluated at $(a, b)$. An arbitrary point of $M_v$ is $c_1(\tau_1)_v + c_2(\tau_2)_v$ where $c_1$ and $c_2$ are real numbers. In the usual coordinates in $\mathbb{R}^3$ this vector has end-point

$$(c_1 + v_1, c_2 + v_2, c_1 \partial f/\partial x + c_2 \partial f/\partial y + v_3)$$

where $v = (v_1, v_2, v_3)$.

(2)  Using 4.9, an immersion $\gamma: \mathbb{R}^2 \to \mathbb{R}^3$ can be used to produce a parametrized surface in $\mathbb{R}^3$ by restriction of the domain of $\gamma$. Thus the tangent space at $v$ to the torus obtained by rotating the curve $\alpha(t) = (\cos t, \sin t + 2)$ (see 4.10(2)) is spanned by

$$(-\sin a, \cos a \cos b, \cos a \sin b)_v$$

**Fig. 4.10.** Tangent plane to a surface

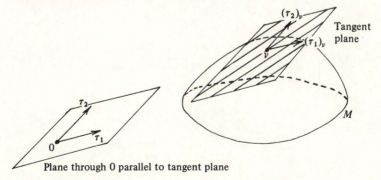

Plane through 0 parallel to tangent plane

and

$$(0, -(\sin a + 2)\sin b, (\sin a + 2)\cos b)_v.$$

For instance when $a=0$ or $\pi$ these vectors span the plane through $v$ parallel to $x=0$. All points with $a=0$ have the same tangent plane, which in the usual coordinates has equation $x=1$ ($a=\pi$ gives $x=-1$) (fig. 4.11). (Can you find those points where the tangent plane is parallel to the plane $y=0$, or to the plane $z=0$?)

(3) The *chain rule for tangent maps* is an immediate consequence of the chain rule for derivatives. Thus, if

$$\mathbb{R}^n, u \xrightarrow{\ f\ } \mathbb{R}^p, v \xrightarrow{\ g\ } \mathbb{R}^q, w$$

are maps, then

$$D(g \circ f)(u) = Dg(v) \circ Df(u)$$

and

$$T(g \circ f)(u) = Tg(v) \circ Tf(u).$$

### 4.23     Exercises

(1) (*i*) Consider a parametrized surface $M$ given by $\phi$ as in 4.7, with $n=2$, $q=1$. Let $\alpha: I \to W$ be a regular curve with $\alpha(0)=w \in W$. Write $\gamma=\phi|W$. Then $\gamma \circ \alpha: I \to \mathbb{R}^3$ will be a regular curve whose image lies in $M$ and which passes through $\gamma(w)=v$ (fig. 4.12). Use the chain rule in (3) above to show that

image $T(\gamma \circ \alpha)(0) \subset$ image $T\gamma(w)$.

**Fig. 4.11.** Torus of revolution

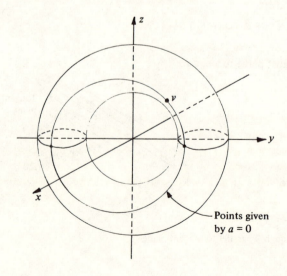

Points given
by $a = 0$

**Fig. 4.12.** Curve in a surface

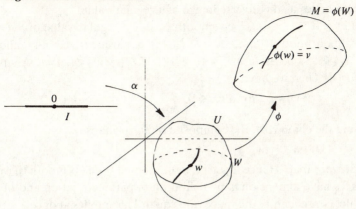

This says that tangent vectors to the space curve $\gamma \circ \alpha$ at $v$ are in the tangent plane to $M$ at $v$.

(*ii*) Suppose everything in (*i*) is given except $\alpha$, and let $y_v$ be any tangent vector to $M$ at $v$, i.e. $y_v = T\gamma(w)(x_w)$ for some $x_w \in \mathbb{R}_w^2$ (recall $W \subset \mathbb{R}^2 \subset \mathbb{R}^3$). This is the same as $y = D\gamma(w)(x)$ for some $x \in \mathbb{R}^2$. Define $\alpha: \mathbb{R}, 0 \to \mathbb{R}^2$, $w$ by $\alpha(t) = w + tx$. Show that $\gamma \circ \alpha$, which is a curve in $M$, has $y_v$ as a tangent vector at $v$. (First show that $T\alpha(0): \mathbb{R}_0^1 \to \mathbb{R}_w^2$ takes 1 to $x_w$, i.e. that $D\alpha(0): \mathbb{R} \to \mathbb{R}^2$ takes 1 to $x$.)

Results (*i*) and (*ii*) show that the tangent space ($=$tangent plane) to $M$ at $v$ consists precisely of tangent vectors to curves in $M$ which pass through $v$.

(2) Generalize (1) to show that the tangent space at $v$ to any parametrized $n$-manifold $M$ in $\mathbb{R}^{n+q}$ consists of tangent vectors to curves in $M$ through $v$.

(3) Let $f: \mathbb{R}^n, u \to \mathbb{R}^p, v$ be a smooth map and let $\alpha: \mathbb{R}, 0 \to \mathbb{R}^n$, $u$ be a (regular) curve in $\mathbb{R}^n$. Use the chain rule to show that $T(f \circ \alpha)(\mathbb{R}_0) \subset \text{image } Tf(u)$. The image of $T(f \circ \alpha)$ is spanned by $T(f \circ \alpha)(0)$. Show that this is the vector

$$\lim_{s \to 0} \frac{f \circ \alpha(s) - f \circ \alpha(0)}{s}.$$

(This is a very convenient way of obtaining tangent vectors in the image of a tangent map $Tf(u)$.)

(4) Let $\alpha$, $\gamma$ be as in 4.10(3), so that the image of $\gamma$ is the tangent developable of the space curve $\alpha$. For $(x_0, y_0) \in I \times \mathbb{R}_+$ there is then a neighbourhood $W_1$ of $(x_0, y_0)$ such that $\gamma(W_1)$ is a parametrized surface in $\mathbb{R}^3$. Show that $D\gamma(x_0, y_0)$ has image spanned by the vectors $\alpha'(x_0)$, $\alpha''(x_0)$. Show that the tangent plane to $\gamma(W_1)$ at $\gamma(x_0, y_0)$ (i.e. the set of end-points of tangent vectors to the surface at $\gamma(x_0, y_0)$) is the set of points

$$\{\alpha(x_0) + \lambda\alpha'(x_0) + \mu\alpha''(x_0): \lambda, \mu \in \mathbb{R}\}.$$

(Not surprisingly, this is independent of $y_0$: varying $y_0$ and keeping $x_0$ fixed just moves $\gamma(x_0, y_0)$ along the tangent line to $\alpha$ at $x_0$.)

Finally we investigate the tangent space to a parametrized manifold which arises as the inverse image of a regular value.

Let $f: \mathbb{R}^{n+q}, v \to \mathbb{R}^q, c$ be smooth, with $c$ a regular value of $f$. Then in a neighbourhood $V$ of $v$, $M = f^{-1}(c)$ is a parametrized $n$-manifold. Let $\phi: U \to \mathbb{R}^{n+q}$ be as in 4.7, with $\phi(W) = f^{-1}(c) \cap V$, $\phi(w) = v$. As usual, let $\gamma = \phi | W$. Then we have

$$f(\gamma(x)) = c \quad \text{for all } x \in W.$$

Using the chain rule, the composite of linear maps

$$Tf(v) \circ T\gamma(w)$$

is the zero map. Hence image$(T\gamma(w)) \subset \text{kernel}(Tf(v))$. But both these vector spaces have dimension $n$, by definition of parametrization and of regular value; consequently the spaces are equal. This implies at once:

### 4.24 Proposition

*The tangent space to $M = f^{-1}(c)$ at $v$ is the kernel of the linear map $Tf(v): \mathbb{R}_v^{n+q} \to \mathbb{R}_c^q$. This equals*

$$\{y_v : y \in \text{kernel } Df(v)\}.$$

*Thus the tangent space passes through $v$ and is parallel to the linear subspace kernel $Df(v)$ of $\mathbb{R}^{n+q}$.* ☐

Note that the kernel is independent of the choice of $\phi$: the only property needed was $f(\gamma(x)) = c$ for all $x$ close to $w$.

For example, let 0 be a regular value of $f: \mathbb{R}^2, v \to \mathbb{R}, 0$. The linear map $Df(v)$ has kernel $\{(\xi, \eta) \in \mathbb{R}^2 : \partial f / \partial x(v)\xi + \partial f / \partial y(v)\eta = 0\}$ and the tangent line to $f^{-1}(0)$ at $v$ consists of vectors $(\xi, \eta)_v$ with $(\xi, \eta) \in \text{kernel } Df(v)$. The end-points of these vectors, i.e. the points $(\xi, \eta) + v$, are therefore the points of $\mathbb{R}^2$ satisfying

$$\frac{\partial f}{\partial x}(v)(x - v_1) + \frac{\partial f}{\partial y}(v)(y - v_2) = 0$$

where $v = (v_1, v_2)$. This is the *equation* of the tangent line in ordinary $(x, y)$ coordinates in $\mathbb{R}^2$.

### 4.25 Exercises

(1) Show similarly that for $f: \mathbb{R}^3, v \to \mathbb{R}, 0$ with 0 a regular value, the equation of the tangent plane to $f^{-1}(0)$ at $v$ is

$$\frac{\partial f}{\partial x}(v)(x - v_1) + \frac{\partial f}{\partial y}(v)(y - v_2) + \frac{\partial f}{\partial z}(v)(z - v_3) = 0.$$

(2) Find the equations of the tangent planes to the surface $x^2 + y^2 - z^2 - 1 = 0$ at points of the form $(v_1, v_2, 0)$. Show that these planes are all parallel to the $z$-axis.

(3) Check that 0 is a regular value of $f: \mathbb{R}^3 \to \mathbb{R}$ where $f(x, y, z) = x^2 + 2y^2 + z^2 - 1$. (Here $f^{-1}(0)$ is an ellipsoid of revolution.) Show that the tangent plane to $f^{-1}(0)$ at $v = (v_1, v_2, v_3)$, $v_1 \neq 0$, has basis $\{(-2v_2, v_1, 0)_v, (-v_3, 0, v_1)_v$. What if $v_1 = 0$? Show that $(v_1, 2v_2, v_3)_v$ is perpendicular to all tangent vectors at $v$ (it is a *normal vector* at $v$).

(4) For $f$ as in 4.24 show that, for $1 \leqslant i \leqslant q$,

$$(\operatorname{grad} f_i(v))_v = \left( \frac{\partial f_i}{\partial x_1}(v), \ldots, \frac{\partial f_i}{\partial x_{n+q}}(v) \right)_v$$

are independent vectors in $\mathbb{R}_v^{n+q}$ perpendicular to all vectors in the tangent space to $f^{-1}(c)$ at $v$. They form a basis for the *normal space* to $f^{-1}(c)$ at $v$; for $q = 1$ the normal space is a line. Apply this to (3) above.

Suppose that we have a surface $M$ in $\mathbb{R}^3$ and look at it from a point $0$ – think of a ball held at arm's length, or a sheet of material folded on a table, or a surface such as a torus (doughnut) made of transparent plastic and viewed from a distance. There will be certain points of $M$ where, from the viewpoint $0$, $M$ appears to curve away from us or to fold in some more complex fashion. At these points the tangent plane to $M$ will pass through $0$, and projecting the points of contact of these planes through $0$ onto a fixed plane $X$ gives the 'apparent contour' of $M$ from $0$: essentially this is the outline as it appears on the retina of the viewing eye (fig. 4.13, left half). We can approximate this situation closely by considering instead those points of $M$ where the tangent plane is parallel to a fixed direction $v$. These points will lie on some curve $C$ in $M$ and, choosing a plane perpendicular to $v$, $C$ projects onto this plane to give the *apparent contour of $M$ in the direction $v$* (fig. 4.13, right half). It is this latter contour which we always consider. For example the apparent contour of the surface $2t^3 + t(1 - 2b) + a = 0$ (see 1.4) in the direction of the $t$-axis ($v = (0, 0, 1)$) is the cusped curve

**Fig. 4.13.** Apparent contours

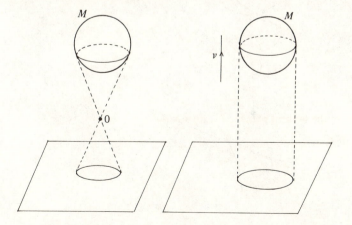

$27a^2 + 2(2b-1)^3 = 0$ (see the remarks following 1.6, and exercise (1) below). Fig. 2.1 shows some apparent contours of a torus surface in $\mathbb{R}^3$.

## 4.26    Exercises

(1) Let $F(t, x, y) = t^3 + xt + y$, $G(t, x, y) = (F(t, x, y), \partial F/\partial t(t, x, y))$. (Compare 1.4, which differs only slightly in notation.) Show that 0 is a regular value of $F$ and $(0, 0)$ is a regular value of $G$. (So $G^{-1}(0, 0)$ is (in a neighbourhood of each point) a parametrized 1-manifold (hence a regular curve) lying on the parametrized surface $F^{-1}(0)$.) Show that the tangent plane to $F^{-1}(0)$ at every point of $G^{-1}(0, 0)$ is 'vertical', i.e. contains the vector parallel to $(1, 0, 0)$. (This is the same as saying kernel $DF(t_0, x_0, y_0)$ contains the vector $(1, 0, 0)$, when $G(t_0, x_0, y_0) = (0, 0)$.) Show that the tangent line to $G^{-1}(0, 0)$ at $(t_0, x_0, y_0)$ is vertical iff $t_0 = 0$ (which implies $x_0 = y_0 = 0$). (In this case the whole of the curve $G^{-1}(0, 0)$ can be parametrized $t \mapsto (t, -3t^2, 2t^3)$ so you can check the latter result by 4.24 or by using this explicit parametrization.) Show that, provided $t_0 \neq 0$, the tangent line to $G^{-1}(0, 0)$ at $(t_0, x_0, y_0)$ projects under $\pi : \mathbb{R}^3 \to \mathbb{R}^2$, $\pi(t, x, y) = (x, y)$, to the line with equation $t_0^3 + t_0 x + y = 0$ (fig. 4.14). (The curve $\pi(G^{-1}(0, 0))$ is the apparent contour of the surface $F^{-1}(0)$ in the direction of the $t$-axis.)

(2) Let $F : \mathbb{R}^3 \to \mathbb{R}$ be smooth, where we use variables $t, x, y$ for $\mathbb{R}^3$. For each $t$, write $F_t : \mathbb{R}^2 \to \mathbb{R}$ for the function defined by $F_t(x, y) = F(t, x, y)$. Suppose that $F$ satisfies the property that, for all $(t, x, y) \in F^{-1}(0)$, either $\partial F/\partial x(t, x, y)$ or $\partial F/\partial y(t, x, y)$ is nonzero. Show that each $F_t$ has 0 as a regular

**Fig. 4.14.** The situation of exercise (1)

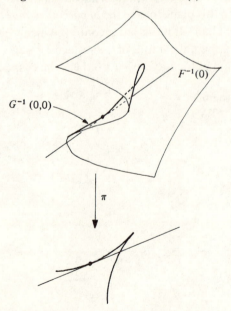

value, so that each $F_t^{-1}(0)$ is (in a neighbourhood of any of its points) a parametrized 1-manifold, and in particular a regular curve. What $F$ gives us is a *family of curves* parametrized by $t$.

Now let $G: \mathbb{R}^3 \to \mathbb{R}^2$ be $G(t, x, y) = (F(t, x, y), \partial F/\partial t(t, x, y))$ and let $(t_0, x_0, y_0) \in G^{-1}(0, 0)$, where $\partial^2 G/\partial t^2(t_0, x_0, y_0) \neq 0$. Show that in a neighbourhood of $(t_0, x_0, y_0)$ in $\mathbb{R}^3$, $G^{-1}(0, 0)$ is a parametrized 1-manifold. Now let $\pi: \mathbb{R}^3 \to \mathbb{R}^2$ be the projection $\pi(t, x, y) = (x, y)$. Show that the image under $\pi$ of the tangent line to the curve $G^{-1}(0, 0)$ at $(t_0, x_0, y_0)$ is precisely the tangent line to the curve $F_{t_0}^{-1}(0)$ at $(x_0, y_0)$. The curve in $\mathbb{R}^2$ obtained by projecting $G^{-1}(0, 0)$ under $\pi$ is called the *envelope* of the family $\{F_t^{-1}(0)\}$. It is also the apparent contour of the surface $F^{-1}(0)$ in the $t$-direction. More of this in chapter 5.

(3) Let $F: \mathbb{R} \times \mathbb{R}^r \to \mathbb{R}$ have 0 as a regular value. Calling the variable in $\mathbb{R}$ by the name $t$, show that the tangent space to $M = F^{-1}(0)$ at $(t_0, a) \in \mathbb{R} \times \mathbb{R}^r$ contains a vector parallel to the $t$-axis iff $\partial F/\partial t(t_0, a) = 0$.

## 4.27 Further exercises

In the light of the work in this chapter on regular values, re-read the early sections of chapter 2 which have to do with contact between a curve $\gamma$ in $\mathbb{R}^n$ and a set (hypersurface) $F^{-1}(0)$ where 0 is a regular value of $F: \mathbb{R}^n \to \mathbb{R}$. The definition can now be phrased: curve and hypersurface have $k$-point contact at $x$, where $\gamma(t) = x$, $F(x) = 0$, iff $F \circ \gamma$ has an $A_{k-1}$ singularity at $t$.

For contact between two curves in $\mathbb{R}^2$ it is desirable to have a more symmetrical definition. Work your way to this via the following suggestions:

(1) A two-variable version of the Hadamard lemma 3.4 says that if $G: \mathbb{R}^2 \to \mathbb{R}$ is smooth and $G(x, 0) = 0$ for all $x$, then $G(x, y) = yG_1(x, y)$ for a smooth $G_1$ and all $(x, y)$ in the domain of $G$. You could prove this in a similar way to 3.4, or look it up, say in Bröcker and Lander (1975) (or just believe it). An easy deduction: if $F(x, f(x)) = 0$ for all $x$ close to $x_0$ (where $F: \mathbb{R}^2$, $(x_0, f(x_0)) \to \mathbb{R}$ and $f: \mathbb{R}$, $x_0 \to \mathbb{R}$ are smooth) then $F(x, y) = (y - f(x))F_1(x, y)$ for a smooth $F_1$ and all $(x, y)$ close to $(x_0, f(x_0))$.

(2) Let $F: \mathbb{R}^2$, $(x_0, y_0) \to \mathbb{R}$ be smooth, and say $\partial F/\partial y(x_0, y_0) \neq 0$. Let $\gamma: \mathbb{R}, t_0 \to \mathbb{R}^2$ be a regular curve, with $\gamma(t_0) = (x_0, y_0)$. Then the order of contact of $\gamma$ and $F^{-1}(0)$ at $(x_0, y_0)$ is measured by the type of $F \circ \gamma$ at $t_0$. Show that, keeping the same contact, $F$ can be replaced by $y - f(x)$ for a unique $f$, and $\gamma$ by $\delta$ where $\delta(x) = (x, g(x))$ for a unique $g$, assuming $\gamma_1'(t_0) \neq 0$. (What if this equals 0?) Use (1), and 2.12(5).

(3) Deduce that the order of contact is $k$ iff the function $f - g$ has type $A_{k-1}$ at $x_0$. Thus whenever the two curves are represented locally by equations $y = f(x)$, $y = g(x)$ (where $g(x_0) = f(x_0)$), the order of contact is measured in this way by the type of singularity of $f - g$ at $x_0$.

(4) Use (3) and 2.28(7) to show that *two curves have k-point contact at a common point iff* (i) *they have the same tangent line there, and* (ii) *the curvature and its derivatives (with respect to arc-length) up to order exactly $k - 3$ have the same value at the point for the one curve as for the other.* This is an

'intrinsic' characterization of contact. (It is enough to consider two curves both tangent to the $x$-axis at $(0, 0)$. They are then locally graphs of functions $y = f(x)$ and $y = g(x)$ and 2.28(7) says that $f$ and $g$ have the same $(k-1)$-jet at 0 iff the above condition (*ii*) on curvatures holds.)

# 5

*Envelopes*

Holmes laughed. 'It is quite a pretty little
problem,' said he.   (*A Scandal in Bohemia*)

What have light caustics, grass fires, gunnery ranges and embroidery in
common? The canny reader, glancing at the title of this chapter, will
immediately answer 'They are all connected with envelopes (whatever
those are)' – and indeed that is exactly right. We suggest that you try to
relate each of the following pictures (fig. 5.1) with one of the above topics.

In each case there are a lot of curves (they might be straight lines),
which represent light rays or trajectories or threads or whatever. These
appear to cluster along another curve, which the eye immediately picks
out; they also touch this other curve. The new curve may look very different from those which gave birth to it; we hope you agree that it can be
very beautiful. The new curve is called the *envelope* of the others.

In chapter 1 we considered all the normals to a given parabola, where
the envelope is a cuspidal cubic curve – see 1.5 and 1.7. There we spread
out the normals to form a surface in $\mathbb{R}^3$; the envelope then appeared as the
contour of this surface when viewed from above. (See figs. 1.2, 1.5.) It is
actually this idea which is formalized in the definition of envelope (5.3),
but in an optional section we also formulate and compare some other
definitions – see 5.8 *et seq.* We then begin to study the local structure of
envelopes, and this study is continued in chapter 7.

## Families and envelopes

Let us begin with a very simple example: a family of circles of
radius 1 centred on the $x_1$-axis in the $(x_1, x_2)$-plane (fig. 5.2). In this case
the envelope visibly consists of the lines $x_2 = 1$ and $x_2 = -1$. Here are
four possible explanations for this.

**Fig. 5.1.** Envelopes

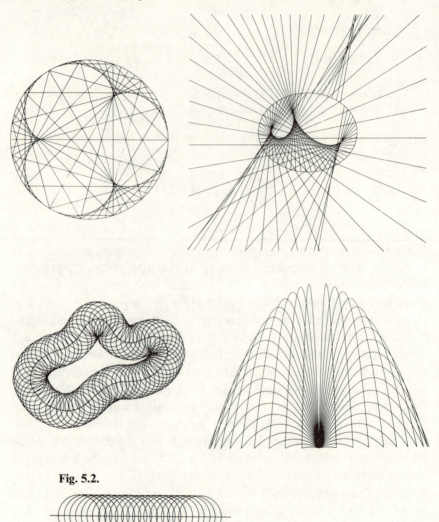

**Fig. 5.2.**

**5.1**     (1) Two circles with nearby centres intersect in two points, which approach the lines $x_2 = \pm 1$ as the circles tend to coincidence. (The old books express this poetically by saying that the lines consist of intersection points of pairs of 'consecutive circles'.)

(2) Each of the lines is tangent to all the circles.

(3) The circles fill up a strip, $|x_2| \leq 1$, whose boundary consists of the two lines.

(4) The family of circles has equation

$$(x_1 - t)^2 + x_2^2 - 1 = 0$$

(i.e. each fixed $t$ gives one of the circles). In $\mathbb{R}^3$ (coordinates $t$, $x_1$, $x_2$) this is a surface (see fig. 5.3). In the direction of the $t$-axis the apparent contour (projection of points of the surface where the tangent plane is parallel to the $t$-axis) consists of the lines $x_2 = \pm 1$ in the $(x_1, x_2)$-plane.

In an optional section below (starting at 5.8) we formalize (1)–(3) for any family of curves and compare the results with (4): it turns out that (1)–(3) always define sets which are *contained in* the set defined by (4). For the present we shall concentrate on the geometry associated with the 'profile' definition (4).

First catch your curves. The way we shall do this is to give them as inverse images of regular values. There is no harm in a little generalization here, so rather than regular values of functions $\mathbb{R}^2 \to \mathbb{R}$ let us take $\mathbb{R}^r \to \mathbb{R}$ ($r = 3$ gives surfaces in $\mathbb{R}^3$).

Suppose that

**5.2**      $F: \mathbb{R} \times \mathbb{R}^r \to \mathbb{R}$

is a smooth map. We use $t$, $x_1, \ldots, x_r$ as coordinates on the left, and think of $F$ as a *family of functions of* $x$, parametrized by $t$. Write $F_t: \mathbb{R}^r \to \mathbb{R}$ for the function $F_t(x) = F(t, x)$; *we suppose that, for each $t$, 0 is a regular value*

**Fig. 5.3.** Envelope as apparent contour of a surface

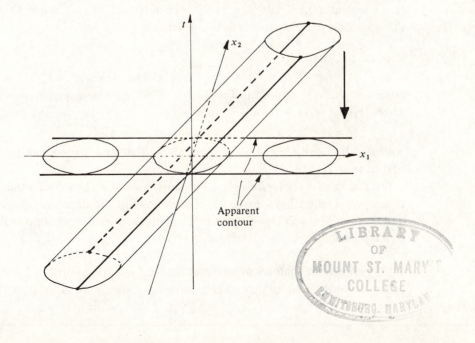

Apparent
contour

LIBRARY OF MOUNT ST. MARY'S COLLEGE EMMITSBURG, MARYLAND

*of* $F_t$ (i.e. whenever $F(t, x) = 0$, we have some $\partial F/\partial x_i$ nonzero). Hence, by 4.16, $C_t = F_t^{-1}(0)$ is a parametrized $r - 1$ manifold in a neighbourhood of each point ($r = 2$: curve; $r = 3$: surface). Note $C_t$ is a curve, not a tangent space!

It is easy to check that 0 is then a regular value of $F$: the Jacobian matrix of $F$ is

$$\left(\frac{\partial F}{\partial t} \frac{\partial F}{\partial x_1} \cdots \frac{\partial F}{\partial x_r}\right) = \left(\frac{\partial F}{\partial t} \frac{\partial F_t}{\partial x_1} \cdots \frac{\partial F_t}{\partial x_r}\right)$$

and by hypothesis the last $r$ entries are never all zero for $(t, x)$ in $F^{-1}(0)$. Hence $F^{-1}(0)$ is (locally) a parametrized $r$-manifold in $\mathbb{R}^{r+1}$. For $r = 2$ it is a surface – the surface obtained by spreading out the plane curves $C_t$, moving $C_t$ to level $t$ in the third dimension.

Thus $F^{-1}(0)$ has a tangent space at $(t, x)$, passing through $(t, x)$ and parallel to kernel $\mathrm{D}F(t, x)$. This tangent space is 'vertical', i.e. parallel to the $t$-axis, iff $(1, 0, \ldots, 0) \in$ kernel $\mathrm{D}F(t, x)$, which is the same as $\partial F/\partial t(t, x) = 0$.

For $r = 2$, the set $F = \partial F/\partial t = 0$ in $\mathbb{R}^3$ is the *fold curve* along which the surface appears to fold when looked at in the $t$-direction. The projection of this to the $(x_1, x_2)$-plane is the envelope. In general, we define:

### 5.3    Definition

The *envelope*, or the *discriminant*, of the family $F$ is the set

$$\mathcal{D} = \mathcal{D}_F = \{x \in \mathbb{R}^r : \text{there exists } t \in \mathbb{R} \text{ with } F(t, x) = \partial F/\partial t(t, x) = 0\}.$$

If $x \in \mathcal{D}$ and $F(t, x) = \partial F/\partial t(t, x) = 0$ then $t$ is said to *correspond* to $x$. (Thus at least one $t$ corresponds to a given $x$.)

### 5.4    Examples

(1)  $F(t, x) = (x_1 - t)^2 + x_2^2 - 1$ ($r = 2$). The curves $C_t = F_t^{-1}(0)$ are circles of radius 1 centred in the $x_1$-axis, as in the first example above. Then $\partial F/\partial t = -2(x_1 - t)$ and

$$\mathcal{D} = \{x \in \mathbb{R}^2 : x_1 = t, x_2^2 = 1 \text{ for some } t\} = \{x \in \mathbb{R}^2 : x_2 = \pm 1\},$$

which is the lines $x_2 = \pm 1$ as expected. Note that $t$ corresponds to $(t, 1)$ and to $(t, -1)$ in $\mathcal{D}$.

(2)  $F(t, x) = 2t^3 + t(1 - 2x_2) - x_1$. The surface $F = 0$ is the folded surface of chapter 1, the curves $F_t = 0$ are the normals to the parabola $x_2 = x_1^2$ and $\mathcal{D} = \{x : 27x_1^2 = 2(2x_2 - 1)^3\}$ as in 1.5. This is the envelope of normals to the parabola.

(3)  **Apparent contours of surfaces**  Instead of starting with a family of curves and spreading them out to form a surface $M$ we can start with

any $F: \mathbb{R}^3 \to \mathbb{R}$, where 0 is a regular value of $F$, and take parallel plane sections of the surface $M = F^{-1}(0)$ to give a family of curves. Still using $t, x_1, x_2$ for coordinates in $\mathbb{R}^3$ consider the sections by planes $t = $ constant. Not all these will necessarily be regular curves, of course, for 0 will not be a regular value of $F_t$ if $\partial F/\partial x_1$ and $\partial F/\partial x_2$ vanish at $(t, x_1, x_2)$ for some $x_1, x_2$. However if this happens then $\partial F/\partial t$ will be *nonzero* so the tangent plane to $M$ at $(t, x_1, x_2)$ will not be 'vertical' (parallel to the $t$-axis) – indeed it will be 'horizontal'. Thus the points $(t, x)$ of $M$ for which the tangent plane is vertical are all regular points of the map $F_t$. Consequently near such points the curves $F_t = 0$ (sections of $M$ by planes perpendicular to the $t$-axis) are regular curves. The envelope of these curves (i.e. the envelope of $F$ restricted to the set of points $(t, x)$ where $\partial F/\partial x_1$ and $\partial F/\partial x_2$ are not both zero) is the apparent contour of $M$ in the $t$-direction. Compare 4.26 and fig. 5.3.

### 5.5 Exercises

(1) Show that the equation of the normal at $(2-t^2, -2t)$ to the parabola $x_2^2 = 4(2-x_1)$ is $F(t, x) = 0$ where $F(t, x) = t^3 + tx_1 + x_2$. Verify that 0 is a regular value of each $F_t$ and that $\mathscr{D} = \{x: 27x_2^2 + 4x_1^3 = 0\}$. What do you get for $F$ and $\mathscr{D}$ by taking the normal at (*i*) $(2-\tfrac{1}{4}t^2, t)$, (*ii*) $(2-t^6, -2t^3)$? (Note that in each case normals and values of $t$ are still in one-to-one correspondence.)

(2) Show that the tangent to $x_2 = x_1^3$ at $(t, t^3)$ is $F(t, x) = 0$ where $F(t, x) = x_2 - 3t^2 x_1 + 2t^3$. Find the envelope of $F$ (what do you *expect* it to be?) Do the same with the envelope of tangents to $x_2 = x_1^2$, calculating the tangent at $(t, t^2)$.

(3) Sketch a few curves $C_t$ and find the envelope for the following $F$; also verify that 0 is a regular value of each $F_t$. (*i*) $x_1 - t^3$, (*ii*) $x_1 - t^2$, (*iii*) $x_2 - (x_1 - t)^3$, (*iv*) $(x_1 - t^3)^2 + x_2^2 - 1$. In each case find all $t$ which correspond to $x \in \mathscr{D}$. Fig. 5.4 shows the surface $F = 0$ in $\mathbb{R}^3$ for the examples (*i*) and (*ii*).

**Fig. 5.4.** The surfaces $x_1 = t^3$ and $x_1 = t^2$

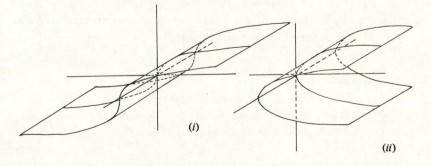

(*i*)

(*ii*)

(4) Let $F(t, x_1, x_2) = t^2 + x_1^2 + x_2^2 - 1$, so $M = F^{-1}(0)$ is the unit sphere. For which $t$ is 0 a regular value of $F_t$? What is the apparent contour of $M$ in the direction of the $t$-axis? (See example (3) above.) Try $F = x_1^2 - x_2^3 - t x_1 + 1$, and sketch some of the sections $t$ = constant of this surface – including the singular ones.

(5) Let $F(t, x) = t^3 + tf + g$ where $f$ and $g$ are functions of $x$. Suppose that $(\partial f/\partial x_1)(\partial g/\partial x_2) - (\partial f/\partial x_2)(\partial g/\partial x_1)$ is never zero. Show that 0 is a regular value of each $F_t$ and that the apparent contour of the surface in the $t$-direction is the curve $27g^2 + 4f^3 = 0$.

As the exercises (1)–(3) show, the envelope depends on $F$ and not just on the curves $C_t$ as sets. When $F(t, x) = x_1 - t$, $\mathscr{D}$ is empty; when $F(t, x) = x_1 - t^3$, $\mathscr{D}$ is the $x_2$-axis, but in each case the $C_t$ are simply all lines parallel to the $x_2$-axis. (So the empty envelope is perhaps the 'expected' one.) There is one simple change in $F$ which *does* preserve $\mathscr{D}$, however.

### 5.6      Proposition

*Let $F$ be as in 5.2 and suppose that $h: \mathbb{R} \to \mathbb{R}$ is a diffeomorphism – in particular $h'$ is never zero. Define $G$ by $G(t, x) = F(h(t), x)$. Then 0 is a regular value of each $G_t$ and $\mathscr{D}_F = \mathscr{D}_G$.* (The same result holds (and the same proof works) if $h$ has domain $U$ such that $h(U)$ contains all $t \in \mathbb{R}$ with $(t, x)$ in the domain of $F$ for some $x$, and $h'$ never vanishes.)

***Proof*** $G$ is the composite $F \circ H$ where $H(t, x) = (h(t), x)$. Applying the chain rule gives

$$\frac{\partial G}{\partial t}(t, x) = \frac{\partial F}{\partial t}(h(t), x)h'(t)$$

and

$$\frac{\partial G}{\partial x_i}(t, x) = \frac{\partial F}{\partial x_i}(h(t), x).$$

It follows that 0 is a regular value of each $G_t$ and $\partial G/\partial t(t, x) = 0$ iff $\partial F/\partial t(h(t), x) = 0$. Hence there exists $t_1$ such that $G = \partial G/\partial t = 0$ at $(t_1, x)$ iff there exists $t_2$ such that $F = \partial F/\partial t = 0$ at $(t_2, x)$. Namely, $t_2 = h(t_1)$ and $t_1 = h^{-1}(t_2)$. $\qquad\square$

Observe that $h(t) = t^3$ does not satisfy the hypothesis of 5.6 at $t = 0$ so on replacing $t$ by $t^3$ in a family $F$ we may have some unexpected arrivals in $\mathscr{D}$ (but $\mathscr{D}$ cannot shrink – can you see why?)

### 5.7      Exercises

(1) Let $F(t, x) = x_1 \cos t + x_2 \sin t - \cos t \sin t$ for all $(t, x) \in \mathbb{R} \times \mathbb{R}^2$. Show that, provided $\sin t \cos t \neq 0$, the line $F(t, x) = 0$ is intercepted by the axes in the $(x_1, x_2)$-plane in a segment of length 1. Show that $t$ corresponds to

$(\sin^3 t, \cos^3 t) \in \mathcal{D}$ and that $\mathcal{D} = \{x : x_1^{\frac{2}{3}} + x_2^{\frac{2}{3}} = 1\}$. This is called an *astroid* – see the left-hand diagram of fig. 5.5. (Can you see a connexion with up-and-over garage doors? The mid-point of the segment actually describes a circle as $t$ varies, and this is also of pivotal importance for the garage doors.)

(2) Let $F(t, x) = x_1(\sin 2t - \sin t) + x_2(\cos t - \cos 2t) - \sin t$, $0 < t < 2\pi$. Show that 0 is a regular value of each $F_t$ and that $F = 0$ is the equation of the chord of the unit circle joining $(\cos t, \sin t)$ to $(\cos 2t, \sin 2t)$. We have: $F = \partial F/\partial t = 0$ iff $x_1 = \frac{1}{3}(2 \cos t + \cos 2t)$, $x_2 = \frac{1}{3}(2 \sin t + \sin 2t)$ (allowing $t = 0$ we find $x_1 = 1$, $x_2$ arbitrary is also a solution). The calculations for this are straightforward but trigonometrically messy. Show that the envelope is a regular curve except for $t = \pi$, corresponding to $x = (-\frac{1}{3}, 0)$. Let $f(t) = F(t, -\frac{1}{3}, 0)$; show that $f$ has an $A_2$ singularity at $t = \pi$. (See the right-hand diagram of fig. 5.5 for a picture of the envelope. Even prettier embroidery is possible if we replace $2t$ by $mt$ in $F$: the envelope has $m - 1$ non-regular points. See fig. 5.1 for the case $m = 4$.)

(3) The trajectory of a particle ejected from the origin with velocity $v$ at an angle $\theta$ to the positive $x_1$-axis is the parabola $x_2 - x_1 \tan \theta + kx_1^2 \sec^2 \theta = 0$, where $k = g/2v^2$ ($g$ = acceleration due to gravity). If you know the necessary dynamics, then you might like to prove this. Letting $F(\theta, x)$ be the left-hand side of this equation (perhaps $t$ should be avoided in a dynamics question) show that the envelope of $F$, where $0 < \theta < \frac{1}{2}\pi$, is the part of the parabola $x_2 = (1/4k) - kx_1^2$ for $x_1 > 0$. (This is the *parabola of safety*: outside it one is safe from attack – provided one knows the value of $v$ and one's enemy is concentrated at the origin. Rotation gives a *paraboloid of safety* for a given $v$ and attacks in three dimensions. See fig. 5.1, bottom right.)

(4) *Envelope of circles of curvature* Let $\gamma : I \to \mathbb{R}^2$ be unit speed. Write down the equation $F(t, x) = 0$ of the circle of curvature at $t$. Show that the envelope of these circles consists of $\gamma(I)$ together with the whole circle of curvature at each vertex of $\gamma$.

**Fig. 5.5.** The astroid and some embroidery

(5) Instead of parametrizing a family of curves by a parameter $t \in \mathbb{R}$ we could parametrize by the points of another curve given by an equation $g(x, y) = 0$ where 0 is a regular value of $g$. Thus the family is given by $F(x, y, u, v) = 0$ where $g(u, v) = 0$. Show that the envelope is obtained by eliminating $u$ and $v$ between these equations and

$$\frac{\partial F}{\partial u}\frac{\partial g}{\partial v} - \frac{\partial F}{\partial v}\frac{\partial g}{\partial u} = 0.$$

Illustrate by considering the family of lines $x/u + y/v = 1$ whose intercepts $u$ and $v$ on the axes satisfy $u^2 + v^2 = 1$ (compare (1) above).

(6) Let $F(t, x) = \|x - \alpha(t)\|^2 - (r(t))^2$ where $\alpha: I \to \mathbb{R}^2$ is unit speed and $r: I \to \mathbb{R}$ is smooth. Thus the discriminant of the family $F$ is the envelope of a family of circles centred on $\alpha(I)$. Assuming $r(t) > 0$ and $|r'(t)| < 1$ for all $t$, show that there are, for each $t$, exactly two points $x$ on the envelope, namely $x = \alpha - rr'T \pm r(1 - r'^2)^{\frac{1}{2}}N$, where $T, N$ refer to $\alpha$. What happens if $r'(t) = 1$ for some $t$? Illustrate with the family of circles centred on the major axis of the ellipse $x_1^2/a^2 + x_2^2/b^2 = 1$ and touching the ellipse. (Suggestion: let $\alpha(t)$ be the point where the normal at $(a \cos t, b \sin t)$ to the ellipse meets the $x_1$-axis. This will not be unit speed, but that does not affect the formula for $x$.)

## Other definitions of envelope

We shall make precise the ideas behind 5.1(1)–(3) and make some attempt to compare these with the definition 5.3 of envelope adopted above. If nothing else, perhaps we shall convince you that (1)–(3) are much harder to work with! We suppose given $F$ as in 5.2, but take $r = 2$ throughout for simplicity.

**5.8**    '*The envelope $E_1$ is the limit of intersections of nearby curves $C_t$*'

Let $E_1 \subset \mathbb{R}^2$ be the set of $x$ for which there exist sequences $(x_n)$ in $\mathbb{R}^2$, $(t_n)$ and $(t'_n)$ in $\mathbb{R}$, as follows. (Note that for once $x_n$ is a point of $\mathbb{R}^2$, say $x_n = (x_{1n}, x_{2n})$.) For all $n$ we require $t_n \neq t'_n$ and $F(t_n, x_n) = F(t'_n, x_n) = 0$, so that $x_n \in C_{t_n} \cap C_{t'_n}$. Also, as $n \to \infty$ we require $t_n \to t$ and $t'_n \to t$ for some $t$, and $x_n \to x$, where $(t, x)$ is in the domain of $F$.

When $n \to \infty$ in $F(t_n, x_n) = 0$ we get $F(t, x) = 0$. Also let $f(t) = F(t, x_n)$ for a fixed $n$ sufficiently large. Then $f(t_n) = f(t'_n) = 0$ so by Rolle's theorem there exists $\tau_n$ between $t_n$ and $t'_n$ with $f'(\tau_n) = 0$. (For all large $n$, the closed interval from $t_n$ to $t'_n$ will be in the domain of $f$, since $t_n$ and $t'_n$ have the same limit $t$.) Hence $\partial F/\partial t(\tau_n, x_n) = 0$, and letting $n \to \infty$, $\partial F/\partial t(t, x) = 0$. Hence:

**5.9**    $E_1 \subset \mathcal{D}$

**5.10    Examples**

(1) For the $F$ of 5.4(1) (circles radius 1 centred on the $x_1$-axis), it is not hard to show that every point of $x_2 = \pm 1$ is in $E_1$. Hence $E_1 = \mathcal{D}$.

(2) For $F(t, x) = x_1 - t^k$ (see 5.5(3)), the curves $C_t$ are disjoint when $k$ is odd, so $E_1$ is empty. For $k \geqslant 2$, $\mathcal{D}$ is the $x_2$-axis so $E_1$ can be strictly smaller than $\mathcal{D}$. For $k = 2$, every point $(0, x_2)$ is the limit of $(1/n^2, x_2) \in C_{1/n}$ and $(1/n^2, x_2) \in C_{-1/n}$, and so belongs to $E_1$. Hence $E_1 = \mathcal{D}$ when $k = 2$. See fig. 5.4: the folding of the surface allows *two* points over each $x$ with $x_1 > 0$.

## 5.11     Exercises

(1) For 3(*iii*) of 5.5 show that $E_1$ is empty. For (*iv*) show that $E_1 = \{x : x_2 = \pm 1\}$. (Hint: replacing $t^3$ by $t$ cannot affect $E_1$.) However, $\mathcal{D}$ is $E_1$ plus the unit circle.

(2) For (2) of 5.5 (tangents to $x_2 = x_1^3$) show that $E_1 = \{x : x_2 = x_1^3 \text{ or } x_2 = 0\}$, which is in fact the same as $\mathcal{D}$. (It is more or less obvious that every point of $x_2 = x_1^3$ is a limit of intersections of nearby tangents, but it may come as a mild surprise that every point $(x_1, 0)$ is too. Show that there exist tangents at points close to $(0, 0)$ intersecting at $(x_1, 1/n)$. You may not be able to find the $t_n$ and $t_n'$ explicitly in terms of $n$.)

(3) *Circles of curvature.* Let $\gamma : I \to \mathbb{R}$ be unit speed and suppose that, for all $t \in I$, $\kappa(t) \neq 0$ and $\kappa'(t) \neq 0$. Show that all the circles of curvature of $\gamma$ are mutually disjoint, so that $E_1$ for this family is *empty*. (Compare 5.7(4) and see also Zeitlin (1981).) Thus if say $\kappa'(t) < 0$ on $I$ (i.e. the radius of curvature $\rho$ is increasing), the circle of curvature at $t_1$ is *inside* the circle of curvature at $t_2$ whenever $t_1 < t_2$. (Hints: this has a quick proof so long as you believe that the straight segment joining the centres of curvature $\varepsilon(t_1)$ and $\varepsilon(t_2)$ ($\varepsilon$ being the evolute) is *shorter* than any other curve, for example the evolute itself. The evolute itself cannot be straight, indeed cannot contain any inflexions at all – see 2.30(3). Using the triangle inequality, we have, for any $x$ on the circle of curvature at $t_1$, $\|x - \varepsilon(t_2)\| \leqslant \rho(t_1) + \|\varepsilon(t_1) - \varepsilon(t_2)\|$. Now use the arc-length formula for the evolute given in 2.30(3).)

(4) Consider a curve with equation $y = f(x)$ where $f$ is smooth and $f(0) = f'(0) = 0$, $f''(0) \neq 0$. Let the tangent at $x = 1/n$ meet the $x$-axis at $(a_n, 0)$. Show that $a_n \to 0$ as $n \to \infty$: hence $(0, 0)$ is a limit of intersections of tangents (at points where $x = 1/n$ and $x = 0$) to the curve. (Harder.) Dropping the condition $f''(0) \neq 0$, show that $(0, 0)$ is still a limit of intersection points of tangents at points where $x = t_n$ and $x = t_n'$, $t_n \neq t_n'$, $t_n$ and $t_n'$ both $\to 0$ as $n \to \infty$. (Why does the choice $t_n = 1/n$, $t_n' = 0$ not now suffice?)

## 5.12     'The envelope $E_2$ is a curve tangent to the $C_t$'

Thus $E_2$ is the set of $x \in \mathbb{R}^2$ for which there exists some regular curve $\gamma : \mathbb{R}, t_0 \to \mathbb{R}^2$ with $\gamma(t_0) = x$ and, for all $t$ in the domain of $\gamma$, (*i*) $\gamma(t) \in C_t$, i.e. $F(t, \gamma(t)) = 0$ and (*ii*) the tangents to $C_t$ and $\gamma$ at $\gamma(t)$ coincide (fig. 5.6).

It follows that image $\gamma \subset E_2$. Also $E_2$ may consist of several pieces, each of which is the image of a regular curve.

Condition (*ii*) can be expressed in various ways, for example

(*ii*)' image $D\gamma(t) = $ kernel $DF_t(\gamma(t))$

**Fig. 5.6.** Curves tangent to the envelope

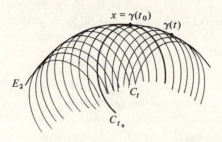

or

$$(ii)'' \quad \frac{\partial F}{\partial x_1}(t, \gamma(t)) \cdot \gamma_1'(t) + \frac{\partial F}{\partial x_2}(t, \gamma(t)) \cdot \gamma_2'(t) = 0$$

where $\gamma(t) = (\gamma_1(t), \gamma_2(t))$.

Note that $(ii)''$ follows from $(ii)'$ since $\gamma'(t)$ is a basis for image $D\gamma(t)$ and $\partial F/\partial x_i(t, x) = \partial F_t/\partial x_i(x)$.

Differentiating condition $(i)$ with respect to $t$ gives

$$\frac{\partial F}{\partial t}(t, \gamma(t)) + \frac{\partial F}{\partial x_1}(t, \gamma(t)) \cdot \gamma_1'(t) + \frac{\partial F}{\partial x_2}(t, \gamma(t)) \cdot \gamma_2'(t) = 0$$

and using $(ii)''$ we find immediately that $\gamma(t) \in \mathcal{D}$. Hence:

**5.13**    $E_2 \subset \mathcal{D}$.

**5.14    Examples**
(1) In the circles example 5.4(1) we can take $\gamma(t) = (t, 1)$ or $(t, -1)$ (all $t \in \mathbb{R}$) to show that both lines $x_2 = \pm 1$ are in $E_2$. Since these make up the whole of $\mathcal{D}$ they make up the whole of $E_2$, too. Replacing $t$ by $t^3$ to give $F(t, x) = (x_1 - t^3)^2 + x_2^2 - 1$ augments $\mathcal{D}$ by the unit circle but no part of this circle is an acceptable $\gamma$. So $E_2 = \{x : x_2 = \pm 1\}$.
(2) For $F(t, x) = x_1 - t^k$, $E_2$ is always empty. ($(ii)''$ implies that $\gamma_1 = $ constant and $(i)$ gives $t = $ constant.) Thus $E_2$ can be smaller than $E_1$ (compare 5.10(2)). For $F(t, x) = x_2 - (x_1 - t)^3$ a possible $\gamma$ is $\gamma(t) = (t, 0)$ so the $x_1$-axis is contained in $E_2$. Since $\mathcal{D}$ is the $x_1$-axis this shows that so is $E_2$. Thus $E_2$ can be larger than $E_1$ (the $C_t$ are all disjoint so $E_1$ is empty).

**5.15    Exercise**
Show that $E_2$ for the 'tangents to $x_2 = x_1^3$' example 5.5(2) is just the curve $x_2 = x_1^3$. For the 'normals to a parabola' example $F(t, x) = t^3 + tx_1 + x_2$ (compare 5.5(1)) show that $E_2 = \{x \in \mathbb{R}^2 : 27x_2^2 + 4x_1^3 = 0 \text{ and } x_1 \neq 0\}$. (Find suitable $\gamma$ for $x_1 < 0$ and $x_2 > 0$ on this cuspidal cubic, use $E_2 \subset \mathcal{D}$ and the fact that $E_2$ cannot contain non-regular curves.)

**5.16**    '*The envelope $E_3$ is the boundary of the region filled by the curves $C_t$.*'

This looks pretty unpromising, but actually does bring out one interesting point. Formally, we define $E_3$ to be the set of $x \in \mathbb{R}^2$ for which (*i*) there exists $t$ with $F(t, x) = 0$, and (*ii*) there exists $\bar{x}$ arbitrarily close to $x$ with $F(t, \bar{x}) \neq 0$ for all $t$.

The interesting point is this. Suppose that we have $(t, x)$ with $F(t, x) = 0$ but $\partial F / \partial t(t, x) \neq 0$. Then the tangent plane to $F^{-1}(0)$ in $\mathbb{R} \times \mathbb{R}^2$ is not vertical, so we would expect $F^{-1}(0)$ not to be folded over $\mathbb{R}^2$ near $(t, x)$, hence to project down to $\mathbb{R}^2$ in a locally one-to-one fashion. In particular for any $\bar{x}$ near $x$ there will exist $\bar{t}$ with $F(\bar{t}, \bar{x}) = 0$, from which it follows that $x \notin E_3$. Likewise, if $x$ is such that $F(t, x)$ is never zero, then $x \notin E_3$, so this will show that $x \notin \mathcal{D}$ implies $x \notin E_3$. Hence:

**5.17**    $E_3 \subset \mathcal{D}$.

The part of the above argument left vague is made precise in 5.20 below.

**5.18**    **Examples**
(1) For the 'circles radius 1 centred on the $x_1$-axis' example 5.4(1), $E_3$ is just the lines $x_2 = \pm 1$. Replacing $t$ by $t^3$, as in (3)(*iv*) of 5.5, clearly makes no difference to $E_3$ (whereas $\mathcal{D}$ is augmented by the unit circle).
(2) For $F(t, x) = x_1 - t^k$, the curves $C_t$ fill up the whole plane for $k$ odd and just the half-plane $x_1 \geqslant 0$ for $k$ even. So $E_3$ is the $x_2$-axis for $k$ even and empty for $k$ odd.

**5.19**    **Exercises**
(1) Show that the normals to $x_2 = x_1^2$ and the tangents to $x_2 = x_1^3$ fill up the plane, so $E_3$ for these sets of lines is empty. (Compare 5.4(2) and 5.5(2).)
(2) Consider the family of circles radius 2 centred on the unit circle. Find a suitable $F$ to describe this family and for this $F$ find $E_1$, $E_2$, $E_3$ and $\mathcal{D}$.
(3) Let $C$ be an *oval*, i.e. a closed curve in the plane with curvature $> 0$ everywhere. Show that $E_3$ for the family of tangent lines to $C$ is precisely $C$ itself.

## Local structure of envelopes

'All this is amusing, though rather elementary,
but I must go back to business, Watson.'
(*A Case of Identity*)

Before investigating the local structure of envelopes let us briefly look at $M = F^{-1}(0)$ and its projection to $\mathbb{R}^r$ (where $F$ is as in 5.2). Near any point $(t_0, x_0)$ of $M$, $M$ is a parametrized $r$-manifold in $\mathbb{R}^{r+1} = \mathbb{R} \times \mathbb{R}^r$, since 0 is a regular value of $F$. As in 4.7, $M$ is locally the image $\gamma(W)$ of an open set $W$ of $\mathbb{R}^r$ by an embedding $\gamma$, the restriction to $W$ of a local diffeomorphism $\phi$. We can study $M$ locally by means of $\gamma$, and the projection

$M \to \mathbb{R}^r$, $(t, x) \mapsto x$, by means of its composite with $\gamma$. We write $\pi$ for the projection $\mathbb{R} \times \mathbb{R}^r \to \mathbb{R}^r$, or for its restriction to $M$, in what follows.

The only properties of $\gamma \colon \mathbb{R}^r$, $w_0 \to \mathbb{R}^{r+1}$ needed are rank $D\gamma(w_0) = r$, $\gamma(w_0) = (t_0, x_0)$ and $F(\gamma(w)) = 0$ for all $w$ close to $w_0$.

### 5.20     Proposition

*With the above notation, $\pi \circ \gamma$ is a local diffeomorphism at $w_0$ if and only if $\partial F/\partial t(t_0, x_0) \neq 0$. (This is precisely the condition for the tangent space to $M$ at $(t_0, x_0)$ not to be vertical.)*

***Proof*** The Jacobian matrix of $\pi$ is an $r \times (r+1)$ matrix with first column zero and the last $r$ columns the identity matrix. Using the chain rule, the Jacobian matrix $A$ of $\pi \circ \gamma$ at $w_0$ is just the last $r$ rows of the Jacobian matrix of $\gamma$ at $w_0$. Using the Inverse Function Theorem we just have to show that $A$ is invertible iff $\partial F/\partial t(t_0, x_0) \neq 0$, i.e. iff $(1, 0, \dots, 0)$ is not in kernel $DF(t_0, x_0)$. Now this kernel is also image $D\gamma(w_0)$ (since $F \circ \gamma = 0$ and kernel and image both have dimension $r$). Writing $v$ for the first row of the Jacobian matrix of $\gamma$ at $w_0$ we have to show

$$\begin{pmatrix} v \\ A \end{pmatrix} \begin{pmatrix} \lambda_1 \\ \vdots \\ \lambda_r \end{pmatrix} = \begin{pmatrix} 1 \\ 0 \\ \vdots \\ 0 \end{pmatrix}$$

has a solution $\lambda$ iff $A$ is singular. But this is an easy exercise in linear algebra. (Remember the whole Jacobian matrix has $r$ independent rows.) $\square$

We can now make the following definitions.

### 5.21     Definitions

A *critical point* of the projection $\pi \colon M \to \mathbb{R}^r$ is a point $(t, x) \in M$ for which $\pi \circ \gamma$ fails to be a local diffeomorphism at the corresponding point $w$. By 5.20 this definition does not depend on the choice of local parametrization $\gamma$. A *critical value* of $\pi \colon M \to \mathbb{R}^r$ is a point $x \in \mathbb{R}^r$ for which some $(t, x)$ is critical.

The set $\Sigma \subset \mathbb{R} \times \mathbb{R}^r$ of critical points of $\pi$ (given by $F = \partial F/\partial t = 0$) is also called the *fold set* of $F$ (the projection $\pi$ has the geometry of the fold, or more complicated behaviour, at points of $\Sigma$). Compare chapter 1, where $M = U^{-1}(0)$ is folded along the curve $\{(t, -4t^3, \frac{1}{2}(1 + 6t^2)\}$ when viewed in the $t$-direction, and 4.26.

The envelope $\mathscr{D}$ of $F \colon \mathbb{R} \times \mathbb{R}^r \rightarrowtail \mathbb{R}$ (0 a regular value of each $F_t$, as usual) is the projection $\pi(\Sigma)$ where $\Sigma$ and $\pi$ are as above. Let $G \colon \mathbb{R} \times \mathbb{R}^r \rightarrowtail \mathbb{R}^2$ be $G(t, x) = (F(t, x), \partial F/\partial t(t, x))$. Then $\Sigma = G^{-1}(0)$. The Jacobian matrix of

$G$ at $(t_0, x_0)$ is

$$\begin{pmatrix} \dfrac{\partial F}{\partial t} & \dfrac{\partial F}{\partial x_1} & \cdots & \dfrac{\partial F}{\partial x_r} \\[2ex] \dfrac{\partial^2 F}{\partial t^2} & \dfrac{\partial^2 F}{\partial x_1 \partial t} & \cdots & \dfrac{\partial^2 F}{\partial x_r \partial t} \end{pmatrix}$$

evaluated at $(t_0, x_0)$. If $(t_0, x_0) \in \Sigma$ then $\partial F/\partial t = 0$; also by assumption some $\partial F/\partial x_i$ is nonzero at $(t_0, x_0)$. Hence:

**5.22    Lemma**

If $\partial^2 F/\partial t^2(t_0, x_0) \neq 0$, *then $G$ is a submersion at $(t_0, x_0)$ so that $\Sigma = G^{-1}(0)$ is locally a parametrized $r - 1$ manifold.*    □

(Of course the same holds when any $2 \times 2$ minor of the above matrix is non-zero.)

**5.23    Example** (See also 4.26(2).)

Consider circles through $(0, \frac{1}{4})$ in the plane with centres on the parabola $x_2 = x_1^2$. Thus for $F$ we take

$$F(t, x) = (x_1 - t)^2 + (x_2 - t^2)^2 - (t^2 + (t^2 - \tfrac{1}{4})^2)$$
$$= x_1^2 - 2tx_1 + x_2^2 - 2t^2 x_2 + \tfrac{1}{2}t^2 - \tfrac{1}{16}.$$

Thus

$$\frac{\partial F}{\partial t}(t, x) = -2x_1 - 4tx_2 + t.$$

Putting $F = \partial F/\partial t = 0$ gives either $x_2 = \frac{1}{4}$, so $x_1 = 0$ and $t$ arbitrary, or $t = 2x_1/(1 - 4x_2)$ and $(x_2 + \frac{1}{4})(x_1^2 + (x_2 - \frac{1}{4})^2) = 0$. Thus $\Sigma = \{(t, 0, \frac{1}{4}): t \in \mathbb{R}\} \cup \{(t, t, -\frac{1}{4}): t \in \mathbb{R}\}$. Also $\partial^2 F/\partial t^2 = -4x_2 + 1$ which is zero on $\Sigma$ just at $(t, 0, \frac{1}{4})$. In $\mathbb{R} \times \mathbb{R}^2$ and $\mathbb{R}^2$ the picture is as in fig. 5.7.

Notice that in this case $\Sigma$ is still a smooth curve even at points where $\partial^2 F/\partial t^2 = 0$, but projection to $\mathbb{R}^2$ gives a bad (isolated) point of the envelope. The smooth part of the envelope is the projection of that part of $\Sigma$ where $\partial^2 F/\partial t^2 \neq 0$.

**5.24    Exercise**

Let $r = 2$ in the discussion before lemma 5.22. Suppose that $\partial^2 F/\partial t^2(t_0, x_0) = 0$ but that nevertheless $G$ is a submersion at $(t_0, x_0)$. Show that the tangent line to $\Sigma$ at $(t_0, x_0)$ is parallel to the $t$-axis. (Compare 4.26(1).) Is $G$ a submersion at each point of the line $x_1 = 0$, $x_2 = \frac{1}{4}$ in 5.23 above?

Thus the structure of the fold locus $\Sigma$ is very nice at a point $(t_0, x_0)$ where $\partial^2 F/\partial t^2 \neq 0$. What about $\mathscr{D}$, which is the projection of $\Sigma$ to $\mathbb{R}^r$? Remember that other values of $t$ besides $t_0$ may correspond to $x_0$: if $t_1$ also corre-

**Fig. 5.7.**

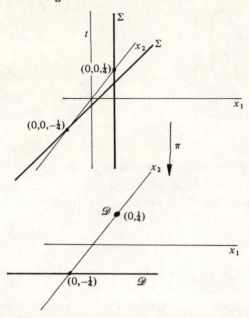

sponds to $x_0$ then a piece of $\Sigma$ near $(t_1, x_0)$ will project to another 'branch' of $\mathscr{D}$ through $x_0$ – or one or both branches might collapse as in 5.23 where *all* points $(t, 0, \frac{1}{4})$ project to $(0, \frac{1}{4}) \in \mathscr{D}$. A less extreme example is indicated in fig. 5.8, where $r = 2$.

We shall concentrate on the part of $\Sigma$ near $(t_0, x_0)$ and investigate the part of $\mathscr{D}$ which is the projection $\pi(\Sigma \cap U)$ for a sufficiently small neighbourhood $U$ of $(t_0, x_0)$ in $\mathbb{R} \times \mathbb{R}^r$.

**Fig. 5.8.** Global structure of $\mathscr{D}$

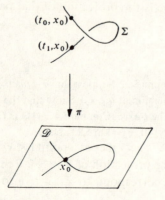

### 5.25 Proposition

*Let $(t_0, x_0) \in \Sigma$ and suppose $\partial^2 F/\partial t^2 \neq 0$ at $(t_0, x_0)$, F being, as usual, as in 5.2. Then there is a neighbourhood U of $(t_0, x_0)$ in $\mathbb{R} \times \mathbb{R}^r$ such that $\pi(\Sigma \cap U)$ is a parametrized $r-1$ manifold in $\mathbb{R}^r$ whose tangent space at $x_0$ coincides with that of $C_{t_0} = F_{t_0}^{-1}(0)$. (Compare 4.26(2).)*

**Proof** Some $\partial F/\partial x_i$ is nonzero at $(t_0, x_0)$; we may suppose that $i = r$. Then by the Implicit Function Theorem applied to $G$ (see 5.22), we can solve for $t$ and $x_r$ as functions of $x_1, \ldots, x_{r-1}$ on $\Sigma \cap U$ for a neighbourhood $U$ of $(t_0, x_0)$. Hence on $\pi(\Sigma \cap U)$, $x_r$ is a smooth function of $x_1, \ldots, x_{r-1}$. Thus writing $x_0 = (x_{01}, \ldots, x_{0r})$ there is a local diffeomorphism $\phi$: $\mathbb{R}^{r-1} \times \mathbb{R}$, $(x_{01}, \ldots, x_{0,r-1}, 0) \to \mathbb{R}^r$, $x_0$ defined by $\phi(y_1, \ldots, y_{r-1}, y_r) = (y_1, \ldots, y_{r-1}, y_r + x_r(y_1, \ldots, y_{r-1}))$.

Clearly then the tangent space to $\pi(\Sigma \cap U)$ at $x_0$ is just the projection by $\pi$ of the tangent space to $\Sigma \cap U$ at $(t_0, x_0)$. The latter is kernel $TG(t_0, x_0)$ and so consists of vectors $(\tau, \xi_1, \ldots, \xi_r)_{(t_0, x_0)}$ with $\Sigma(\partial F/\partial x_i)(t_0, x_0)\xi_i = 0$ and

$$(\partial^2 F/\partial t^2)(t_0, x_0)\tau + \sum(\partial^2 F/\partial t \partial x_i)(t_0, x_0)\xi_i = 0,$$

sums being from 1 to $r$. Since $\partial^2 F/\partial t^2(t_0, x_0) \neq 0$ the projection to $\mathbb{R}^r_{x_0}$ consists precisely of vectors $(\xi_1, \ldots, \xi_r)_{x_0}$ satisfying the first of these equations, and that is the tangent space to $F_{t_0}^{-1}(0)$ at $x_0$. $\quad\square$

Returning to the usual case $r = 2$, this shows that the part of the envelope corresponding to $(t, x) \in \Sigma$ near $(t_0, x_0)$, where $\partial^2 F/\partial t^2(t_0, x_0) \neq 0$, is a regular curve tangent to the curve $C_{t_0}$ of the family at $x_0$ (fig. 5.9).

### 5.26 Definition

*A point of regression is a point $x \in \mathcal{D}_F$ (F as in 5.2) for which there exists $t$ with $(t, x) \in \Sigma$ and $\partial^2 F/\partial t^2 = 0$ at $(t, x)$. The remaining points of $\mathcal{D}_F$ are called regular.*

The very name 'point of regression' suggests something bad, degeneracy if not decadence, and indeed envelopes have been quietly dropped from texts on differential geometry just because of their embarrassingly regres-

**Fig. 5.9.** Curve tangent to envelope at a regular point

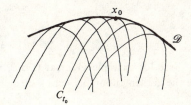

sive tendencies. From the point of view of singularity theory points of regression are of great interest: often they are calling our attention to telling and subtle geometry in the situation being studied. More of this in chapter 7.

### 5.27    Exercises

(1) Look back through the examples of this chapter and find the points of regression. Where possible relate these to some feature of the envelope, such as a cusp or isolated point, where the envelope is not smooth.

(2) Let $F: \mathbb{R} \times \mathbb{R}^2$, $(t_0, x_0) \to \mathbb{R}$ be as usual, and suppose that $F = \partial F/\partial t = 0$ at $(t_0, x_0)$ and, at $(t_0, x_0)$, the matrix

$$\begin{pmatrix} \dfrac{\partial F}{\partial x_1} & \dfrac{\partial F}{\partial x_2} \\[2ex] \dfrac{\partial^2 F}{\partial x_1 \partial t} & \dfrac{\partial^2 F}{\partial x_2 \partial t} \end{pmatrix}$$

is nonsingular. Show that the curve in $\mathbb{R} \times \mathbb{R}^2$ given by $F = \partial F/\partial t = 0$ (i.e. the fold set) can be parametrized, close to $(t_0, x_0)$, by $t$, i.e. that on the curve $x$ is a smooth function $x = x(t)$. Show that the curve in $\mathbb{R}^2$ given by $t \mapsto x(t)$ is regular at $t_0$ iff $\partial^2 F/\partial t^2 (t_0, x_0) \neq 0$. (Thus when the envelope can be parametrized locally by $t$, points of regression are just the points where this parametrization fails to be regular.)

(3) This problem is about $E_1$ (see 5.8). Let $F$ be as in 5.2, with $r = 2$, and suppose $F = \partial F/\partial t = 0$, $\partial^2 F/\partial t^2 \neq 0$ at $(t_0, a_1, a_2)$. Suppose that also $\partial F/\partial x_2 \neq 0$ there (recall some $\partial F/\partial x_i$ is nonzero). Use the Implicit Function Theorem to show that there is a smooth function $f: \mathbb{R}, t_0 \to \mathbb{R}$ satisfying $F(t, a_1, f(t)) = 0$ for all $t$ close to $t_0$ (so that $t \mapsto (t, a_1, f(t))$ locally parametrizes the section of $F^{-1}(0)$ by the plane $x_1 = a_1$). Show that $f$ has a strict local minimum or maximum at $t_0$ and deduce $(a_1, a_2) \in E_1$. Thus all regular points of $\mathscr{D}$ certainly belong to $E_1$. For the example $F(t, x) = (x_1 - t^3)^2 + x_2^2 - 1$, where $\mathscr{D}$ is $\{x: x_2 = \pm 1 \text{ or } x_1^2 + x_2^2 = 1\}$ and $E_1 = \{x: x_2 = \pm 1\}$, does $E_1$ consist precisely of the regular points of $\mathscr{D}$? How about $F(t, x) = x_1 - t^2$ or $x_1 - t^3$?

(4) Let $F(t, x_1, x_2) = x_1 \cos t + x_2 \sin t - f(t)$ where $f$ is smooth. Thus $F = 0$ defines a family of lines in the plane. Find the envelope of this family in the form $x = x(t)$. Show that points of regression are given by $f(t) + f''(t) = 0$, and that the arc-length of the envelope from $t_0$ to $t$ (where $f + f''$ is nonzero on $[t_0, t]$) is

$$\pm \left( f'(t) - f'(t_0) + \int_{t_0}^t f(t)\,dt \right).$$

(This is Legendre's formula – see for example Goursat (1917), p. 430.) Compare 5.7(1), and evaluate the arc-length for $f(t) = \cos t \sin t$.

## Examples of envelopes

The remainder of this chapter consists of examples in which we find the
regular points and points of regression.

### 5.28 Evolutes and caustics

Let $\gamma: I \to \mathbb{R}^2$ be a unit speed plane curve. We consider the *envelope
of normals* of this curve.

The envelope is of some physical interest. Think of light rays starting simultaneously at all points of $\gamma(I)$ and propagating down the normals. Then $\gamma(I)$ is called an
'initial wavefront' and the curve along which the light tends to focus is called the
'caustic'. Focusing takes place where nearby normals intersect, i.e. on the set $E_1$ of
5.8, which actually coincides with the usual envelope $\mathscr{D}$ here (see 5.29(2) below).
See fig. 1.5 for a picture of the normals to a parabola focusing along the envelope.
(We adhere patriotically to the classical, Newtonian optics in which light corpuscles
bounce happily around the universe with no hint of wave motion. However the
reader can be assured that we reject the theory that light rays emanate from the eye.)

The equation of the normal corresponding to $t$ is $(x - \gamma(t)) \cdot \gamma'(t) = 0$ so
we take for $F: I \times \mathbb{R}^2 \to \mathbb{R}$ the function $F(t, x) = (x - \gamma(t)) \cdot \gamma'(t)$. Then 0 is a
regular value of each $F_t$, i.e. we never have $F = \partial F/\partial x_1 = \partial F/\partial x_2 = 0$.
We have: $\partial F/\partial t = (x - \gamma) \cdot T' - T \cdot T = (x - \gamma) \cdot \kappa N - 1$ (compare 2.28(3) and
2.27) and $F = \partial F/\partial t = 0$ gives $\kappa \neq 0$ and $x = \gamma + N/\kappa$, the centre of curvature
of $\gamma$ corresponding to $t$.

Thus $x$ belongs to the envelope precisely when $x$ is a centre of curvature:
*the envelope of normals is precisely the locus of centres of curvature – the
evolute – of the curve.* Here is an informal reason for this: a circle focuses
perfectly at its centre, so for any curve focusing occurs along the locus of
centres of best-fitting circles, that is along the evolute (are you convinced,
or at least illuminated?).

What of the points of regression? Now

$$\partial^2 F/\partial t^2 = (x - \gamma) \cdot \kappa' N - (x - \gamma) \cdot \kappa^2 T$$

so $F = \partial F/\partial t = \partial^2 F/\partial t^2 = 0$ gives $x = \gamma + N/\kappa$ where $\kappa' = 0$. Thus $x$ is a
point of regression iff it is the centre of curvature at a vertex, and away
from such points the envelope – hence the evolute of $\gamma$ – is locally a
regular curve. Compare 2.27, 2.30(3). Later (in chapter 7) we shall be able
to say something about the envelope at a point of regression.

### 5.29 Exercises

(1) For a unit speed $\gamma: I \to \mathbb{R}^2$ define

$F(t, x) = (x - \gamma(t)) \cdot N(t)$

(compare 2.28(2)) so that $C_t = F_t^{-1}(0)$ is the tangent at $t$. For the envelope

$\mathscr{D}_F$ of tangents, show

(i) If $\kappa(t)$ is never zero, then there are no points of regression;

(ii) $\mathscr{D}_F$ consists of $\gamma(I)$ and the tangent lines at all points where $\kappa$ is zero;

(iii) If $\kappa(t_0)=0$ then $\gamma(t_0) \in \mathscr{D}_F$ is a point of regression;

(iv) If $\kappa(t_0)=\kappa'(t_0)=0$ then every point of the tangent line at $t_0$ is a point of regression.

Since neither $\mathscr{D}$ nor the conditions $\kappa=0$, $\kappa'=0$ are affected by reparametrization (compare 5.6) the same results hold for any (regular) curve $\gamma$. Apply them to $\gamma(t)=(t, t^3)$ and $\gamma(t)=(t, t^4)$ (compare 5.5(2)).

(2) (This is about $E_1$, as in 5.8.) For the envelope $\mathscr{D}$ of normals as in 5.28 above show that $E_1 = \mathscr{D}$. (It is only necessary to show that the centre of curvature corresponding to $t_0 \in I$ where $\kappa(t_0) \neq 0$ is a limit of intersections of nearby normals; in fact you can take the normals corresponding to $t_0$ and $t_0+1/n$, as $n\to\infty$. It might help to choose the curve in standard position with $t_0=0$, $\gamma(0)=(0, 0)$, $\gamma'(0)=(1, 0)$.)

(3) Let $\gamma: I \to \mathbb{R}^2$ be a unit speed plane curve and suppose $\kappa(t)$ is never zero. Define $\Gamma: I \times \mathbb{R} \to \mathbb{R} \times \mathbb{R}^2$ by $\Gamma(t, u)=(t, \gamma(t)+uN(t))$. Show that $\Gamma$ is an immersion (i.e. rank $D\Gamma(t, u)=2$ everywhere) and also injective. By 4.9 the image $M$ of $\Gamma$ is (near any point) a parametrized 2-manifold. Write down a basis for the tangent plane to this 2-manifold at the point corresponding to $(t, u)$. Find the points of $M$ where the tangent plane contains a vector parallel to $(1, 0, 0)$. Show that the set $\Sigma$ of such points is the image of a regular curve in $\mathbb{R}^3 = \mathbb{R} \times \mathbb{R}^2$, parametrized $t \mapsto (t, \gamma(t)+ N(t)/\kappa(t))$. Find the points of this curve where the tangent line is parallel to $(1, 0, 0)$. Writing $\pi: \mathbb{R} \times \mathbb{R}^2 \to \mathbb{R}^2$ for $\pi(x, y, z)=(y, z)$, show that $\pi(\Sigma)$ is the evolute of $\gamma$.

**5.30    Parallels of plane curves** (compare 2.30(2)).

Let $\gamma: I \to \mathbb{R}^2$ be unit speed; we consider the envelope of circles centred on $\gamma(I)$ and having a fixed radius $r>0$. Thus we take $F(t, x) =(x-\gamma(t)) \cdot (x-\gamma(t))-r^2$. Then $0$ is a regular value of each $F_t$, $\partial F/\partial t =2(x-\gamma(t)) \cdot T(t)$ and $\partial F/\partial t=0$ gives $x-\gamma(t)=\lambda N(t)$ for some $\lambda \in \mathbb{R}$. Then $F=0$ gives $\lambda=\pm r$ so

$$\mathscr{D}_F=\{\gamma(t) \pm rN(t): t \in I\},$$

which is precisely the parallels of $\gamma$ at distance $\pm r$: the curves obtained by moving down the normals a fixed distance in one direction or the other.

Again this example has an optical interpretation. Viewing $\gamma(I)$ as an initial wavefront the parallel at distance $r$ is the wavefront after a time $r$/(speed of light). For pyromaniacs, the example has further interest. Imagine $\gamma(I)$ to be a trail of inflammable substance on a uniform flat dry grassland (this really sounds like applied mathematics!) Igniting $\gamma(I)$ the fire will spread so that at any instant the burnt area is the union of disks of fixed radius centred on $\gamma(I)$. Thus the boundary separating burnt and unburnt grass is part of the envelope of circles considered above – in fact the part called $E_3$ in 5.16.

To find the points of regression we use

$$\partial^2 F/\partial t^2 = 2(-T \cdot T + (x-\gamma)\cdot \kappa N).$$

Thus $F = \partial F/\partial t = \partial^2 F/\partial t^2 = 0$ gives $x = \gamma + N/\kappa$ (and $\kappa \neq 0$) so $x$ is a point of regression iff it is the centre of curvature of $\gamma$ at $t$, and $r = \pm(1/\kappa)$, whichever is $>0$. (This result holds whether $\gamma$ is unit speed or not.) See fig. 2.21 which shows some parallels of a parabola. Figure 5.10 shows some parallels of a cubic oval.

As $r$ increases, the points of regression on the parallels sweep out the evolute of $\gamma$. In the diagrams these points of regression usually show up as cusps: they are points where the parallel may fail to be a regular curve. More information in 7.12.

### 5.31    Exercise

Parametrizing a parallel to $\gamma$ by $\delta(t) = \gamma(t) + r N(t)$, show that the tangents to $\gamma$ and the parallel at points with parameter $t$ are *parallel lines*, provided $\delta(t)$ is not a point of regression. Equivalently, $\gamma$ and $\delta$ have the *same* normal line at these points. Suppose now that $r = 1/\kappa(t_0)$ and that $\kappa'(t_0) \neq 0$ so that $\kappa$ is monotonic on some neighbourhood of $t_0$ and no value $t \neq t_0$ close to $t_0$ gives a point of regression. Show that the limiting direction of the tangent

**Fig. 5.10.** Parallels of $4y^2 = 9(x - x^3)$, and circles radius 0.4

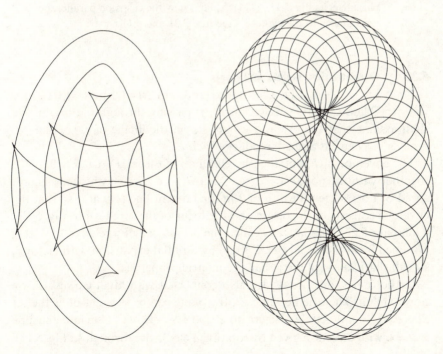

**Fig. 5.11.** Parallel through a centre of curvature

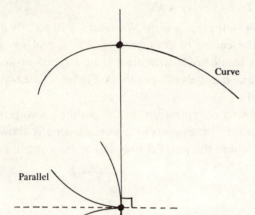

Curve

Parallel

Evolute

to $\delta$, as $t \to t_0$, is along the normal to the evolute of $\gamma$ at the point with parameter $t_0$ (fig. 5.11). (This explains why the cusps on parallels point at right angles to the evolute – see figs. 2.21 and 5.11.)

**5.32     Orthotomics, pedals and duals**

Let $\gamma: I \to \mathbb{R}^2$ be unit speed and $F(t, x) = (\gamma(t) - x) \cdot (\gamma(t) - x) - \gamma(t) \cdot \gamma(t)$. Then $C_t$ is the circle centre $\gamma(t)$ passing through the origin. We shall assume that the origin does not lie on the image of $\gamma$, i.e. that $\gamma(t)$ is never 0. Then 0 is a regular value of each $F_t$.

The envelope of $F$, obtained by eliminating $t$ from $F = \partial F/\partial t = 0$, is $\{0\}$ together with the points $x = 2(\gamma(t) \cdot N(t))N(t)$, $t \in I$. This (whether $\gamma$ is unit speed or not) is the reflexion of 0 in the tangent line to $\gamma$ at $t$, so that the envelope consists of the origin and the locus of these reflexions (fig. 5.12). (Compare 5.23.) The latter is called the *orthotomic* of $\gamma$ relative to 0. It is obtained from the pedal (see 2.30(1)) by a radial expansion with factor 2, so will share many of the pedal's geometric properties.

Orthotomics and pedals are also connected with 'dual curves', as we now explain. The oriented lines in the plane can be thought of as the set $S^1 \times \mathbb{R}$ ($S^1 =$ unit circle), where to $(a, \lambda) \in S^1 \times \mathbb{R}$ we associate the line $x \cdot a = \lambda$, with direction given by rotating $a$ clockwise through $\frac{1}{2}\pi$ (fig. 5.13).

**Fig. 5.12.** Orthotomic

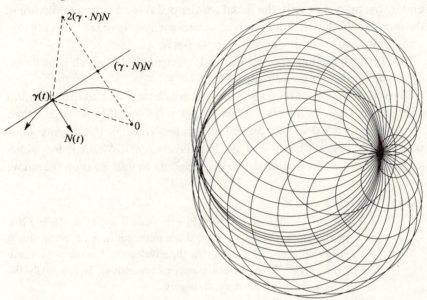

Thus $(a, \lambda)$ and $(-a, -\lambda)$ give the same line with opposite orientations.

Given a plane curve $\gamma$ we can associate to each (oriented) tangent line the corresponding point of $S^1 \times \mathbb{R}$, namely $(N(t), \gamma(t) \cdot N(t))$, and this subset of $S^1 \times \mathbb{R}$ is called the *dual curve* to $\gamma$. Now $S^1 \times \mathbb{R}$ can be thought of as a cylinder but it is more convenient to think of $(a, \lambda)$ as polar coordinates in $\mathbb{R}^2$, i.e. to project $S^1 \times \mathbb{R}$ to $\mathbb{R}^2$ by $(a, \lambda) \mapsto \lambda a$. This map is a 'double cover' since $(a, \lambda)$ and $(-a, -\lambda)$ go to the same thing, and it is 'ramified at 0',

**Fig. 5.13.** Oriented lines in the plane

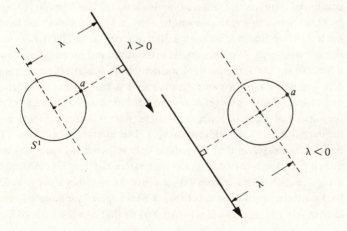

i.e. the whole circle $\{(a, 0): a \in S^1\}$ goes to 0. The image of the dual curve under this map is exactly the pedal, so the pedal is an accurate picture of the dual except that (*i*) it does not tell us the orientation of the tangents to $\gamma$, and (*ii*) tangents to $\gamma$ through 0 all go to $0 \in \mathbb{R}^2$.

The orthotomic has another (optical) interpretation, which is followed up in the exercises and following 7.17.

To find the points of regression $x$ of the envelope $F$ we add the condition $\partial^2 F / \partial t^2 = 0$, i.e. $-2\kappa N \cdot x = 0$. Thus $(\gamma \cdot N)\kappa = 0$. Now if $\gamma \cdot N = 0$ at $t$ then $x = 0$ (note $\gamma \cdot N = 0$ iff the tangent at $t$ passes through the origin). Thus points of regression are 0 and the points $x = 2(\gamma(t) \cdot N(t))N(t)$ for which $\kappa(t) = 0$: points of the orthotomic corresponding to *inflexions* on the curve.

### 5.33    Exercises

(1) Let $F: \mathbb{R} \times (\mathbb{R}^2 - \{0\}) \to \mathbb{R}$ be $F(t, r, \theta) = \gamma(t) \cdot (\cos \theta, \sin \theta) - r$. Here $\gamma$ is a unit speed plane curve and $(r, \theta)$, $r > 0$, are polar coordinates. Show that 0 is a regular value of each $F_t$ and that the envelope of $F$ is the pedal curve minus the origin. What are the points of regression? Is this really the construction for the orthotomic in disguise?

(2) With $\gamma$ a unit speed plane curve not passing through the origin, and $\delta(t) = 2(\gamma \cdot N)N$ the orthotomic relative to 0, show the following.
(*i*) $\frac{1}{2}\delta'(t) = -\kappa((\gamma \cdot T)N + (\gamma \cdot N)T)$ (at $t$ of course);
(*ii*) If $\delta'(t) = 0$ then $\kappa(t) = 0$.
Now assume $\kappa > 0$ at all points of $\gamma$, so that by (*ii*) $\delta$ is regular.
(*iii*) The normal to $\delta$ at $t$ has direction $(\gamma \cdot T)T - (\gamma \cdot N)N$;
(*iv*) The normal to $\delta$ at $t$ passes through $\gamma(t)$ – i.e. its direction is that of $\gamma(t) - \delta(t)$.

Consider a point source of light at 0, the light from 0 being reflected from the curve $\gamma(I)$ according to the usual rule that the reflected ray and the incident ray make equal angles on opposite sides of the normal. Then by congruent triangles the reflected ray is along the line from $\delta(t)$ to $\gamma(t)$, and by (2)(*iv*) above this is the *normal* to $\delta$ (fig. 5.14). It follows that light rays having the orthotomic as their initial wavefront (i.e. light rays starting simultaneously at all points on the orthotomic and propagating down the normals) are the same as light incident from 0 and reflected from $\gamma(I)$. The *caustic by reflexion* of $\gamma$ relative to 0 is the evolute (caustic) of the orthotomic. Caustics by reflexion can be seen clearly as a bright curve on the surface of a cup of coffee in sunlight: here 0 is very far away and the incident light rays are almost parallel. The mathematical caustic by reflexion, illustrated in fig. 5.15 (where only reflected rays are drawn) contains a part that is invisible on the coffee since the light rays creating it are diverging after reflexion from the *outside* of the cup: they must be produced backwards to reveal what nature hides. See chapter 7 for more information.

(3) Let $\gamma(t) = (t, t^3 + 1)$, $\delta(t) = 2(\gamma(t) \cdot N(t))N(t)$ so that $\delta$ is the orthotomic of $\gamma$,

**Fig. 5.14.** Reflected ray

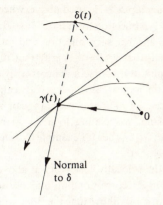

**Fig. 5.15.** Caustic by reflexion of a circle, parallel light

despite the fact that $\gamma$ is not unit speed. Show that the 3-jet with constant of $\delta$ at 0 is $(-6t^2, 2-4t^3)$. What does the curve given by $t \mapsto (-6t^2, 2-4t^3)$ look like near $t=0$? (It will approximate $\delta$ for small $t$.) Note that $\gamma$ has an inflexion at $t=0$.

(4) Repeat exercise (3) for $\gamma(t)=(t, t^4+1)$, going to the 4-jet with constant of $\delta$ at 0, and $\gamma(t)=(t-1, t^3)$, going to the 5-jet. Note that in the latter case, the origin is on the inflexional tangent to $\gamma$.

(5) Let $\mathscr{L}$ be the set of unoriented lines in $\mathbb{R}^2$ not through the origin, and let $l \in \mathscr{L}$. We can associate with $l$ a point $\phi(l) \in \mathbb{R}^2$ by taking any equation

$a_1 x_1 + a_2 x_2 = \lambda$ for $l$ and defining $\phi(l) = \lambda a/(a_1^2 + a_2^2)$. (Note that $\lambda \neq 0$ and $(a_1, a_2) = a \neq (0, 0)$.) This is the same as the map $\mathscr{L} \to \mathbb{R}^2$ used above, where $a$ was taken to have length $l$. Show that $\phi$ is a bijection from $\mathscr{L}$ to $\mathbb{R}^2 - \{0\}$. Show also that $\psi(l) = (a_1/\lambda, a_2/\lambda)$ defines another such bijection and find a smooth map $h: \mathbb{R}^2 - \{0\} \to \mathbb{R}^2 - \{0\}$ such that $h \circ \phi = \psi$. Verify that the Jacobian of $h$ is nonsingular at every point. (Of course $h$ is necessarily a bijection since $\phi$ and $\psi$ are; it is a 'global diffeomorphism' and is classically known as inversion in the unit circle. The set of *all* lines in the plane can be parametrized by the 'real projective plane', associating to $a \cdot x = \lambda$ the point with "homogeneous coordinates" $(a_1 : a_2 : \lambda)$. See for example Coxeter (1969).)

(6) The map $\psi$ in (5) can be used to examine duals locally, just as $\phi$ was above. Take $\gamma(t) = (t, t^3 + 1)$ and find the curve in $\mathbb{R}^2$ corresponding under $\psi$ to the set of tangents to $\gamma$ at points close to $t = 0$. Show that the 3-jet with constant of this curve at 0 is $(3t^2, 1 + 2t^3)$. Investigate locally the set of *normals* to $\gamma(t) = (t, t^2 + 1)$ close to $t = 0$, also using $\psi$. Does the set of normals, regarded in this way as a 'dual curve' in $\mathbb{R}^2$, look like the evolute of $\gamma$?

(7) A variant of $\psi$ as in (5) is as follows. Suppose that $l$ has equation $a_1 x_1 + x_2 = \lambda$ (so that $l$ is not parallel to the $x_2$-axis). Then associate to $l$ the point $(a_1, -\lambda) \in \mathbb{R}^2$. This is as good as $\phi$ or $\psi$ in (5) for studying duals close to points where the tangent line is not parallel to the $x_2$-axis. Let $\gamma(t) = (t, f(t))$ where $f(0) = f'(0) = 0$, $f''(0) \neq 0$. Write down the equation of the tangent line to $\gamma$ at $t$ and show that, using the above map $l \mapsto (a_1, -\lambda)$, the tangent line corresponds to $\delta(t) = (-f'(t), -f(t) + tf'(t)) \in \mathbb{R}^2$, so that (near $t = 0$), $\delta$ is the dual curve of $\gamma$. Show that the tangent line to the *dual* curve $\delta$ at $t$ corresponds to the point $(t, f(t)) \in \mathbb{R}^2$: the dual of the dual is identified with the original curve.

(8) Suppose that the dual $\delta$ in (7) is given close to $(0, 0)$ by the equation $x_2 = F(x_1)$ (compare 2.31). Show that $F(0) = F'(0) = 0$. Writing $f(t) = a_2 t^2 + \cdots + a_k t^k + t^{k+1} f_1(t)$ $(a_2 \neq 0)$ and $F(u) = b_2 u^2 + \cdots + b_k u^k + u^{k+1} F_1(t)$ (see 3.4; the $a_i$ and $b_i$ are real numbers here), show that $a_2, \ldots, a_k$ determine $b_2, \ldots, b_k$ and vice versa. Thus *the $k$-jet of a curve $x_2 = f(x_1)$ determines the $k$-jet of its dual $x_2 = F(x_1)$ and vice versa.*

(9) Let $\gamma(t) = (t, f(t))$, $\beta(t) = (t, g(t))$ $(f, g$ smooth), where $f(0) = f'(0) = g(0) = g'(0) = 0$, so that $\gamma$ and $\beta$ touch at the origin, and $f''(0)$ $g''(0)$ are both nonzero. Assume the result of 4.27 that $\gamma$ and $\beta$ have $k$-point contact at $(0, 0)$ if and only if $f - g$ has type $A_{k-1}$ at 0. Use (8) to show that *the order of contact of the curves $\gamma$ and $\beta$ at $(0, 0)$ is equal to the order of contact of their duals at $(0, 0)$.*

## 5.34    The tangent developable of a space curve    (compare 4.10(3))

Let $\gamma: I \to \mathbb{R}^3$ be a unit speed space curve which has $\kappa(t) \neq 0$ for all $t \in I$. Thus at each point there is a well-defined tangent $T(t)$, principal normal $N(t)$ and binormal $B(t)$ (see 2.33). The osculating plane at $t$ is spanned by $T(t)$ and $N(t)$, and here we consider the envelope of these planes, so we use

$$F(t, x) = (x - \gamma(t)) \cdot B(t).$$

Thus

$$\partial F/\partial t = (x - \gamma) \cdot (-\tau N) - T \cdot B$$
$$= -(x - \gamma) \cdot \tau N$$

and $F = \partial F/\partial t = 0$ gives $x = \gamma(t) + \lambda T(t) + \mu N(t)$ where $\tau(t) = 0$ or $\mu = 0$. Hence the envelope consists of the entire osculating planes at the points of zero torsion together with all the tangent lines to $\gamma$. (This is where the name 'tangent developable' comes from.)

Points of regression: $\partial^2 F/\partial t^2 = 0$ gives $(x - \gamma) \cdot (-\tau' N + \tau \kappa T - \tau^2 B) = 0$.

(i) When $\tau \neq 0$ we find $\lambda = \mu = 0$, so $x = \gamma(t)$.

(ii) When $\tau = 0$, $\tau' \neq 0$ we find $\mu = 0$, so $x = \gamma(t) + \lambda T(t)$.

(iii) When $\tau = \tau' = 0$, we find no condition, so $x = \gamma(t) + \lambda T(t) + \mu N(t)$. Hence every point of the curve, every point of the tangent line at a point where $\tau = 0$ and every point of the osculating plane where $\tau = \tau' = 0$, is a point of regression on the envelope. Away from these, the tangent developable is (locally) a parametrized 2-manifold. See fig. 7.5 for a drawing of 'half' the tangent developable (the part with $\lambda \geq 0$) for a helix, where $\tau$ is never zero.

## 5.35 Exercise

Investigate similarly the envelope of normal planes to a unit speed space curve, by taking $F(t, x) = (x - \gamma(t)) \cdot T(t)$. This is the analogue of the evolute for a space curve. The calculations in 2.34 may help, since $\frac{1}{2} f'$ is just $-F$.

## 5.36 Discriminants of polynomials

The general monic polynomial of degree 3, $t^3 + \lambda t^2 + \mu t + \nu$, can be reduced by replacing $t$ by $t - \frac{1}{3}\lambda$ (Tschirnhaus's transformation) to the form $F(t, x_1, x_2) = t^3 + x_1 t + x_2$. Cubic polynomials in this form can be identified with $\mathbb{R}^2$ (coordinates $x_1, x_2$) and for a fixed $t$ the set $\{x : F(t, x) = 0\}$ is the straight line corresponding to polynomials having $t$ as a root. What is the envelope of these lines? (Certainly 0 is a regular value of each $F_t$, since $\partial F/\partial x_2$ is *never* zero.) The envelope, $\mathcal{D} = \{x : \text{for some } t, F = \partial F/\partial t = 0 \text{ at } (t, x)\}$ is just the set of polynomials having a repeated root, for a repeated root of a polynomial is shared with its derivative. Calculation gives $\mathcal{D} = \{x : 4x_1^3 + 27x_2^2 = 0\}$, with $t$ corresponding to $(-3t^2, 2t^3) \in \mathcal{D}$. The polynomial $4x_1^3 + 27x_2^2$ is traditionally known as the *discriminant* of the cubic: its vanishing is the condition for a repeated root and its sign distinguishes one from three real roots (fig. 5.16). (Of course the calculation above has been done at least once before in this book, for example in chapter 1.)

**Fig. 5.16.** Roots of a reduced cubic

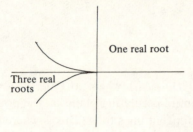

One real root

Three real
roots

In chapter 1 (see 1.4) we used $U(t, a, b) = 2t^3 + t(1 - 2b) - a$, so that $F(t, x) = \frac{1}{2}G(t, \frac{1}{2}(1 - 2x_1), -2x_2)$. The set $F = 0$ is a linear transform of the surface $U = 0$ in fig. 1.2, folded along a curve which consists of points $(t, -3t^2, 2t^3)$, projecting to $\mathscr{D}$ in $\mathbb{R}^2$.

Let us repeat the procedure for the reduced quartic

$$F(t, x) = t^4 + x_1 t^2 + x_2 t + x_3.$$

Fixing $t$ gives a plane $F_t(x) = 0$ in $\mathbb{R}^3$ (note $\partial F / \partial x_3$ is never zero). We can ask for the envelope of these planes. As before, regarding $\mathbb{R}^3$ as the space of reduced (and monic) polynomials of degree 4, the envelope is just the set of such polynomials having a repeated real root. Now $F = \partial F / \partial t = 0$ gives $x_2 = -(4t^3 + 2x_1 t)$, $x_3 = 3t^4 + x_1 t^2$, so as $x_1$ and $t$ vary, the point

$$(x_1, -4t^3 - 2x_1 t, 3t^4 + x_1 t^2)$$

describes the discriminant surface $\mathscr{D}$. The set $M = F^{-1}(0)$ and the fold set $\Sigma = \{x : F = \partial F / \partial t = 0\}$ are now in $\mathbb{R}^4$, so we cannot draw them, but $\mathscr{D}$ can be drawn. Points of regression are given by $\partial^2 F / \partial t^2 = 0$, which just means that $t$ is a triple root of $F = 0$, so we can expect $\mathscr{D}$ to fail to be a parametrized 2-manifold along the curve $\{(-6t^2, 8t^3, -3t^4) : t \in \mathbb{R}\}$. Elsewhere it definitely will be smooth.

Taking sections of $\mathscr{D}$ by planes $x_1 = \lambda = $ constant, we obtain pictures (fig. 5.17) which fit together to form the surface in fig. 5.18. This is called the *swallowtail* or *dovetail* surface, according to taste. (The French is *queue d'aronde*, which means dovetail as in carpentry, but aronde is also an archaic word for swallow.) The curve $\{(-6t^2, 8t^3, -3t^4)\}$ is the cuspidal edge of the swallowtail; note that there is also a line of self-intersection. This corresponds to quartic polynomials with roots of the form $(a, a, b, b)$. In fact for fixed $x_1 = \lambda < 0$ the pattern of roots of $F(t, x) = 0$ is shown in fig. 5.19.

For the quintic $t^5 + x_1 t^3 + x_2 t^2 + x_3 t + x_4$ we obtain the *butterfly*, which lives in $\mathbb{R}^4$ and so can only be drawn in 3-dimensional sections, say for $x_1$ fixed. The swallowtail and butterfly arise in applications with a control

**Fig. 5.17.** Sections of a swallowtail

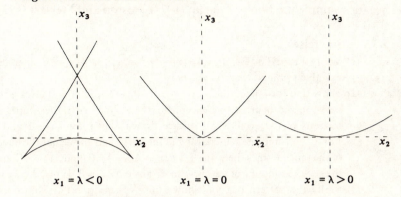

$x_1 = \lambda < 0$      $x_1 = \lambda = 0$      $x_1 = \lambda > 0$

**Fig. 5.18.** The swallowtail surface. (This computer picture was done for us by David Fidal.)

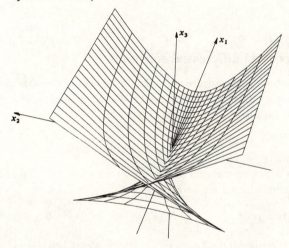

**Fig. 5.19.** Number of real roots of a reduced quartic

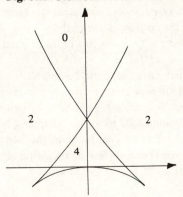

space $\mathbb{R}^3$ or $\mathbb{R}^4$, just as the cusp in chapter 1 arose with a control space $\mathbb{R}^2$. See for example the books Zeeman (1977), Poston and Stewart (1978).

**5.37    Exercises**

(1) Verify that $\gamma : \mathbb{R}^2 \to \mathbb{R}^3$, $\gamma(t, x_1) = (x_1, -4t^3 - 2x_1 t, 3t^4 + x_1 t^2)$ is an immersion exactly at points where $x_1 \neq -6t^2$.

(2) Show that for $\lambda$ fixed the map $t \mapsto (\lambda, -4t^3 - 2\lambda t, 3t^4 + \lambda t^2)$ gives a regular curve in the plane $x_1 = \lambda$ except for $t^2 = -\lambda/6$. Verify the shapes of the sections of the swallowtail surface drawn above, for $\lambda = -1$ and $\lambda = 1$. For $\lambda = -1$ find the self-crossing of the curve and show that the equations of the tangent lines there are $\pm 2\sqrt{2}x_2 + 4x_3 - 1 = 0$ (and $x_1 = \lambda$). Show that the self-crossings form the curve in $\mathbb{R}^3$ given by $x_2 = 0$, $4x_3 = x_1^2$, $x_1 \leqslant 0$.

(3) Verify the pattern of roots shown for $\lambda < 0$ in fig. 5.19. (The roots vary continuously with $x$, so you may assume that within one connected region, where the roots are all distinct, the pattern is the same for every $x$.) Do the same for $\lambda > 0$.

## Envelopes and differential equations

One reason why envelopes were studied in days of old was their connexion with first order ordinary differential equations (ODEs). Envelopes arose as so-called 'singular' solutions; here is an example. See also p. 149.

**5.38    Example**

Consider the one-parameter family of parabolas $y = (x - c)^2$. By differentiating this equation and eliminating $c$ we obtain

$$\left(\frac{dy}{dx}\right)^2 - 4y = 0.$$

Clearly (and not surprisingly) the functions $y = (x - c)^2$ are all solutions of this first order ODE, but so also is the function $y = 0$. Geometrically the reason is clear: the line $y = 0$ is the envelope of the family of parabolas, and since the tangent to the envelope at each point coincides with the tangent to one of the solution curves the envelope is also a solution. Since the line $y = 0$ was not part of the original family of curves it is termed a 'singular solution'.

We shall not venture far into what is in fact a rather murky backwater of our subject. To quote R. P. Agnew ((1960), p 104), 'It may be true that there is no subject in mathematics upon which more unintelligible nonsense has been written than that of families of curves'. (The reader is recommended to look at chapter 4 of Agnew's book, especially p 116, for a lively indictment of what he terms the 'mathematical medicine men' of the none too distant past.) Nevertheless no discussion of envelopes would be complete without some mention of Clairaut's equation.

**5.39     Example**

Let $f$ be a smooth function of one variable and consider the ODE

$$y = \frac{dy}{dx} x + f\left(\frac{dy}{dx}\right)$$

which is known as the Clairaut equation. Differentiating, we find

$$\frac{dy}{dx} = \frac{dy}{dx} + \frac{d^2y}{dx^2}\left(x + f'\left(\frac{dy}{dx}\right)\right),$$

so that either $d^2y/dx^2 = 0$ or $x = -f'(dy/dx)$. In the first case $y = ax + b$ say and substituting in the Clairaut equation we find $b = f(a)$. In the second case we find $y = -f'(dy/dx) \cdot dy/dx + f(dy/dx)$. In fact writing $x = -f'(a)$, $y = -af'(a) + f(a)$, so that $(x, y)$ is parametrized by $a$, we find $dy/dx = a$ and $(x, y)$ satisfies the Clairaut equation for all $a$. Clearly this solution is the envelope of the family of straight line solutions $y = ax + f(a)$.

**5.40     Exercises**

(1) Solve the Clairaut equation $y = (dy/dx)x + c/(dy/dx)$.

(2) Find a curve, other than a straight line, such that the tangent at any point determines, together with the $x$ and $y$ axes, a triangle of area $\frac{1}{2}c^2$.

(3) There are uncountably many other solutions (differentiable, but not smooth) of the equation in 5.38. Draw a picture and piece together parts of the solutions given to obtain others valid for all $x$.

**5.41     Projects**

Find out what you can about the following envelopes, (*i*)–(*iv*). (*i*) The orthotomic of a space curve, given as the envelope of spheres centred on the curve and passing through the origin. (*ii*) The envelope of a family of circles of variable radius centred on a given curve (compare 5.7(6)). (*iii*) The envelope of a family of spheres of variable radius centred on a given space curve. This last has connexions with the cyclides of Dupin, which occur for example in Eisenhart (1909). (*iv*) The envelope of a family of circles centred on a given plane curve and touching a given line. This is the 'orthotomic from infinity' of the plane curve in the direction normal to the line, and corresponds to parallel incident light in this normal direction (compare the remarks after 5.33(2)).

(*v*) Investigate the local form of the orthotomic of a plane curve in more detail, after the manner of 5.33(3) and (4).

(*vi*) Investigate the relationship between the envelope of a family of plane curves at a regular point $x$ and the curve $C_t$, where $t$ corresponds to $x$. They touch, by 5.25, but could they have higher contact? Could they cross?

# 6

## *Unfoldings*

'It sounds high-flown and absurd,
but consider the facts!'
(*The Naval Treaty*)

Let $F$ be a family of functions containing the function $f$. For example, $F$ might be the family of potential functions of chapter 1 (see 1.1), which depended on two parameters $a$ and $b$, or it might be the family of distance-squared functions on a curve in $\mathbb{R}^n$, where there are $n$ parameters given by the coordinates of a point $u \in \mathbb{R}^n$ (see 2.13). Changing the parameters a bit will 'perturb' $f$ into a nearby function. In fig. 6.1 are the graphs of some perturbations of $f(t) = t^5$ (the axes are not drawn, but the $t$-axis is in each case horizontal).

The label $k$ at the point of the graph where $t = t_0$ indicates that the function has a turning point which is an $A_k$ singularity at $t = t_0$, or equivalently that the tangent line there is horizontal and has $(k+1)$-point contact with the curve. Notice that sometimes two turning points can be on the same *level*, i.e. the function has the same value there, and sometimes, as with two $A_2$s, this is not possible.

A family of functions containing $f$ is also called an *unfolding* of $f$: the family unfolds to reveal all these functions which are $f$'s close relations. Certain unfoldings contain *all* functions close to $f$ (in a precise sense). For $f(t) = t^5$ such a 'universal unfolding' is $F(t, x_1, x_2, x_3) = t^5 + x_1 t^3 + x_2 t^2 + x_3 t$ – all the pictures below can be obtained by choosing suitable constants $x_i$. Notice that 'higher terms' $t^6, t^7, \ldots$ are not needed. (Can you

**Fig. 6.1.** Perturbations of $t^5$

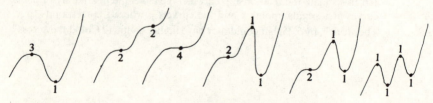

draw the pictures for functions close to $t^6$? What does the graph of a polynomial of degree 6 close to $t^6$ look like?)

It is not in fact this feature of universal unfoldings which is most crucial for us in this book. We are more concerned with two other properties of such families. First, 'almost all' families of functions are universal unfoldings (strictly, 'versal' unfoldings) of each function in the family. Consequently one can expect these unfoldings to arise in virtually any situation where we are studying families of functions. Secondly these unfoldings are in a certain sense unique, i.e. they depend only on the function they are unfolding. Thus one expects the same models, describing the geometry of the unfolding, to arise in many (almost all) situations. Two geometric consequences of the second of these properties are proved in 6.14p and 6.14 below. A rather vague justification of the other property is made at the end of this chapter (6.21). A more precise discussion uses results from chapter 8 (see 8.20(5), (6)).

The existence of these unfoldings which contain all nearby functions is the central result of this chapter (6.6p and 6.6). We sketch the steps of the argument here but unfortunately a complete proof requires the use of a very technical piece of mathematics called the Malgrange Preparation Theorem, and we prefer to omit it. Proofs of more general results can be found in, for example, Bröcker and Lander (1975). Instead, in chapter 10, we prove the corresponding result in which *smooth* functions are replaced by *analytic* functions (given by convergent power series). Thus for analytic functions and curves our treatment is complete.

Unfoldings will be used in the next chapter to study some interesting geometrical situations, such as envelopes in a neighbourhood of a point of regression.

## Unfoldings and (p)versal unfoldings

Let $F: \mathbb{R} \times \mathbb{R}^r, (t_0, x_0) \to \mathbb{R}$ be a smooth function. We can naturally regard $F$ as an $r$-parameter family of functions $F_x: \mathbb{R}, t_0 \to \mathbb{R}$ or as a 1-parameter family of functions $F_t: \mathbb{R}^r, x_0 \to \mathbb{R}$. In chapter 5 we inclined to the latter view but here we favour the former. We write

$$f = F_{x_0}: \mathbb{R}, t_0 \to \mathbb{R}$$

for the function $f(t) = F(t, x_0)$.

### 6.1 Definition

$F$ as above is called an *r-parameter unfolding* of $f$.

Recall from chapter 3 that every function $f: \mathbb{R}, t_0 \to \mathbb{R}$ having an $A_k$ singularity at $t_0$ can be reduced, in a neighbourhood of $t_0$, to one of the forms $g(t) = \pm t^{k+1}$. That is,

$$f(t) = g(h(t)) + c$$

for a local diffeomorphism $h: \mathbb{R}, t_0 \to \mathbb{R}, 0$ and a constant $c$. Suppose we have an unfolding $F$ of $f$. What can this be reduced to, by a suitable change of variable? We sketch the steps of the reduction, but postpone discussion of the difficult ones until chapter 10.

*Step 1* First we can reduce $f$ as above, i.e. define

$$\bar{F}(t, x) = F(h^{-1}(t), x) - c,$$

so that $\bar{F}(t, x_0) = g(t)$ for all $t$ close to 0. Let us take $g(t) = + t^{k+1}$ in what follows.

*Step 2* By a procedure similar to that used in proving the classification theorem 3.3 we can write

$$\bar{F}(t, x) = b_0(x) + b_1(x)t + \cdots + b_{k+1}(x)t^{k+1} + t^{k+2}\theta(t, x)$$

for uniquely determined functions $b_i: \mathbb{R}^r, x_0 \to \mathbb{R}$ and $\theta: \mathbb{R} \times \mathbb{R}^r, (0, x_0) \to \mathbb{R}$. Since $\bar{F}(t, x_0) = t^{k+1}$ we have $b_i(x_0) = 0$ for $0 \le i \le k$, and $b_{k+1}(x_0) = 1$. (The procedure of 3.3 will produce this formula for each fixed $x$, but actually the resulting functions $b_i$ will be *smooth*.)

*Step 3* For this step we must put on our seven-league boots. In chapter 3 we were able to eliminate the 'higher terms' from an $A_k$ function. Here we can actually use a parametrized change of variable to eliminate $t^{k+2}\theta(t, x)$, and also to reduce $b_{k+1}(x)$ to 1 and (as in the Tschirnhaus transformation) remove the term in $t^k$. All of this says that for *new* functions $b_i$ and $c$ we have

$$\bar{F}(a_1(t, x), x) = c(x) + b_1(x)t + \cdots + b_{k-1}(x)t^{k-1} + t^{k+1},$$

where $a_1: \mathbb{R} \times \mathbb{R}^r, (0, x_0) \to \mathbb{R}$ is smooth, $a_1(t, x_0) = t$ for all $t$ close to 0 and $b_i(x_0) = 0$ for $i = 1, \ldots, k-1$, $c(x_0) = 0$.

**6.2p**

> Let $G: \mathbb{R} \times \mathbb{R}^{k-1} \to \mathbb{R}$ be defined by
>
> $$G(t, u) = u_1 t + u_2 t^2 + \cdots + u_{k-1}t^{k-1} + t^{k+1}.$$
>
> Then steps 1 to 3 say that, for $a_1$, $b$ and $c$ as above we have
>
> $$\bar{F}(a_1(t, x), x) = G(t, b(x)) + c(x).$$

*Step 4* Let $a_1 \times id: \mathbb{R} \times \mathbb{R}^r, (0, x_0) \to \mathbb{R} \times \mathbb{R}^r$ be $(a_1 \times id)(t, x) = (a_1(t, x), x)$. Then it is easy to check that $a_1 \times id$ is a local diffeomorphism at $(0, x_0)$, using $a_1(t, x_0) = t$. Define $a: \mathbb{R} \times \mathbb{R}^r, (0, x_0) \to \mathbb{R}$ by $a \times id = (a_1 \times id)^{-1}$, so that $a(t, x) = \lambda$ iff $a_1(\lambda, x) = t$. We can write

**6.3p**    $$\bar{F}(t, x) = G(a(t, x), b(x)) + c(x)$$
> where $a(t, x_0) = t$ (for all $t$ close to 0), $b(x_0) = 0$, $c(x_0) = 0$.

This completes the reduction of $\bar{F}$ to a 'standard form' $G$ via changes of variable given by $a$, $b$, $c$. *We shall assume* 6.3p *in what follows, and discuss the matter further – in particular give a proof for analytic functions – in chapter 10.*

Notice that for each fixed $x$ close to $x_0$, the change of variables is just an $\mathcal{R}$-equivalence between $\bar{F}_x$ at 0 and $G_{b(x)}$ at $a(0, x)$. For $t \mapsto a(t, x)$ will be a local diffeomorphism at 0 for $x$ close to $x_0$, since $a(t, x_0) = t$ implies $\partial a / \partial t(0, x_0) \neq 0$ so that by continuity $\partial a / \partial t(t, x) \neq 0$ for all $(t, x)$ close to $(0, x_0)$.

The transformations appearing in 6.3p are the kind we shall allow when changing one unfolding of a function (in this case $t^{k+1}$) into another unfolding of the same function. If the unfoldings were of merely $\mathcal{R}$-equivalent functions (as with $f$ and $t^{k+1}$ above) we would drop the requirement that $a(t, x_0) = t$, replacing it with $\partial a / \partial t(t_0, x_0) \neq 0$. Before writing down the formal definitions, here is a word about our terminology.

### 6.4    Terminology

For each fixed $x$, $c(x)$ is just an additive constant in 6.3p. By allowing changes of variable where we can add a constant to our functions we are saying that a function $f(t)$ is not significantly different from $f(t) + c$. This is appropriate when we are in a 'potential function' situation, as with $V(t)$ in chapter 1 or with distance-squared and height functions in chapter 2, for there is always a certain arbitrariness in choosing the zero level of potential. But in other circumstances, as with the functions defining curves in chapter 5, it is not appropriate. There, we consider say $F(t, x_1, x_2) = 0$, and $F(t, x_1, x_2) = c$ is not the same thing at all. We are forced to give slightly different allowable changes of variable (omitting $c(x)$) in the latter situation, and this makes all the results very marginally different.

Since we want to cover both cases in this book there is nothing for it but to use terminology that distinguishes them. Rather than write 'potential' every time we shall prefix words by (p), apologizing for the odd appearance. We shall also append p to reference numbers. The other, non-p case, starts after 6.19 below.

### 6.5p    Definition

Let $G: \mathbb{R} \times \mathbb{R}^s$, $(t_0, y_0) \to \mathbb{R}$ be an $s$-parameter unfolding of the function $g = G_{y_0}$. Let

$a: \mathbb{R} \times \mathbb{R}^r$, $(t_0, x_0) \to \mathbb{R}$    where $a(t, x_0) = t$ close to $t_0$

$b: \mathbb{R}^r$, $x_0 \to \mathbb{R}^s$, $y_0$

$c: \mathbb{R}^r$, $x_0 \to \mathbb{R}$

be smooth. Then the unfolding $F: \mathbb{R} \times \mathbb{R}^r, (t_0, x_0) \to \mathbb{R}$ defined by

$$F(t, x) = G(a(t, x), b(x)) + c(x)$$

is said to be *(p)induced from* $G$. (Note that $F$ is an unfolding of $f$, where $f(t) = g(t) + c(x_0)$, which is $g$ up to an additive constant.)

If $G$ is such that every unfolding of $g$ is (p)induced from $G$ then $G$ is called a *(p)versal unfolding of* $g$ at $t_0$. (The quaint word 'versal' is 'universal' without the suggestion of uniqueness.)

Hence the steps sketched above lead to the following central result.

**6.6p**     *The unfolding* $G: \mathbb{R} \times \mathbb{R}^{k-1}, (0, 0) \to \mathbb{R}$, *given by*

$$G(t, x) = \pm t^{k+1} + x_1 t + x_2 t^2 + \cdots + x_{k-1} t^{k-1}$$

*is a (p)versal unfolding of* $g(t) = \pm t^{k+1}$ *at* $t_0 = 0$. □

This unfolding is also called *miniversal* since the number of unfolding parameters is minimal – see 6.10p below.

**6.7p     Example**

Let $G(t, x_1) = t^3 + x_1 t$ and $H(t, x_1, x_2) = t^3 + x_1 t + x_2 t^2$ be unfoldings of the function $t^3$. Then it is easy to (p)induce $G$ from $H$: we merely write

$$G(t, x_1) = H(t, x_1, 0),$$

i.e. $H(a(t, x_1), b(x_1)) + c(x_1)$ where $a(t, x_1) = t$, $b(x_1) = (x_1, 0)$ and $c(x_1) = 0$.

We can also (p)induce $H$ from $G$ using the Tschirnhaus transformation to get rid of the $t^2$ term: thus, by calculation

$$H(t - \tfrac{1}{3}x_2, x_1, x_2) = t^3 + t(x_1 - \tfrac{1}{3}x_2^2) + \tfrac{2}{27}x_2^3 - \tfrac{1}{3}x_1 x_2$$

$$= G(t, x_1 - \tfrac{1}{3}x_2^2) + \tfrac{2}{27}x_2^3 - \tfrac{1}{3}x_1 x_2.$$

Hence

$$H(t, x_1, x_2) = G(t + \tfrac{1}{3}x_2, x_1 - \tfrac{1}{3}x_2^2) + \tfrac{2}{27}x_2^3 - \tfrac{1}{3}x_1 x_2.$$

In this case $a(t, x_1, x_2) = t + \tfrac{1}{3}x_2$: note that $a(t, 0, 0) = t$ as is required by the definition. Here, as in the reduction leading to 6.6p, it is more straightforward to find a formula $H(a_1(t, x), x) = G(t, b(x)) + c(x)$ and then to find $a(t, x)$ by the rule $a_1(\lambda, x) = t$ iff $a(t, x) = \lambda$, making $H(t, x) = G(a(t, x), b(x)) + c(x)$.

It is equally easy to (p)induce $H(t, x) = t^3 + u_1(x)t + u_2(x)t^2$ from $G$, where $u_1$ and $u_2$ are smooth functions of $x = (x_1, \ldots, x_r)$.

We can also add 'higher terms': $H(t, x) = t^3 + t^3\theta(t, x)$ where $\theta(t, 0) = 0$, can be (p) induced from $G$ by

$$H(t, x) = G(t(1 + \theta(t, x))^{\frac{1}{3}}, 0, 0), (t, x) \text{ close to } (0, 0).$$

The *hard* part of 6.6p (even for $k = 2$) comes when we add both 'lower' and 'higher' terms to $t^3$, for example $H(t, x_1, x_2) = t^3 + x_1 t^2 + x_2 t^4$. But 6.6p does assure us that we *can* (p)induce this from $G$. One could find explicit $a$, $b$ and $c$ (well, more or less explicit!) from the method in the first part of chapter 10.

## 6.8p    Exercises

(1) Suppose that $F$, $G$, $H$ are all unfoldings of $g$ (at $t_0$) and that $F$ is (p)induced from $G$ and $G$ is (p)induced from $H$. Show that $F$ is (p)induced from $H$. (This is just a matter of combining say $F(t, x) = G(a(t, x), b(x)) + c(x)$ and $G(u, y) = H(A(u, y), B(y)) + C(y)$, but there is one condition to check – see 6.5p.) Hence show that, if $G$ is (p)versal and $G$ is (p)induced from $H$ then $H$ is (p)versal.

(2) Let $H: \mathbb{R} \times \mathbb{R}^k, (0,0) \to \mathbb{R}$ be $H(t, x) = t^{k+1} + x_1 t + x_2 t^2 + \cdots + x_k t^k$. Show that $H$ is a (p)versal unfolding of $H_0$ ($H_0(t) = t^{k+1}$) at 0. (Induce $G$ as in 6.6p from $H$ and use (1).)

(3) Show how to (p) induce $H$ as in (2) from $G$ as in 6.6p. (This must be possible, since $G$ is (p)versal. Use Tschirnhaus, as in 6.7p above.)

(4) Let $1 \leqslant l \leqslant k$. Find points $x$ arbitrarily close to $0 \in \mathbb{R}^k$ such that $H_x$ ($H$ as in (2)) has a singularity $A_l$ at $t = 0$. (Try most of the $x_i$ equal to zero.) Show also that $H_x$ cannot have type $A_{>k}$ at any $t$. Using the fact that $H$ is a (p)versal unfolding of $t^{k+1}$ at 0, show the following holds for every (p)versal unfolding $F: \mathbb{R} \times \mathbb{R}^r, (0, x_0) \to \mathbb{R}$ of $t^{k+1}$ at 0. Given $l$ as above and a neighbourhood $U$ of $(0, x_0)$ in $\mathbb{R} \times \mathbb{R}^r$, there exists $(t, x) \in U$ such that $F_x$ has a singularity of type $A_l$ at $t$. This shows that every (p)versal unfolding $F$ of $t^{k+1}$ at 0 'contains $A_{<k}$ singularities arbitrarily close to $t^{k+1}$'. Show also that $U$ can be chosen such that $F_x$ never has type $A_{>k}$ at $t$, for $(t, x) \in U$.

(5) Consider the 1-parameter unfolding of $t^6$ ($t_0 = 0$) given by $F(t, x) = (t - x)^2 t^2 (t + x)^2$ ($(t, x)$ close to $(0, 0)$). Show that, for $x \neq 0$, $F_x$ has two $A_1$ singularities (at $t_1$ and $t_2$ say) at one level (i.e. $F_x(t_1) = F_x(t_2)$) and three $A_1$ singularities at a second level. Deduce that if $G: \mathbb{R} \times \mathbb{R}^r, (0, 0) \to \mathbb{R}$ is any (p)versal unfolding of $t^6$ at 0 then there exist $t_1, t_2, t_3, t_4, t_5 \in \mathbb{R}$ and $y \in \mathbb{R}^r$, all arbitrarily close to 0, such that $G_y$ has $A_1$ singularities at all of the $t_i$, those at $t_1, t_2$ being at one level and those at $t_3, t_4, t_5$ being at a second level.

(6) Establish step 2 of the reduction process given above. (This requires some work on Hadamard's lemma.)

(7) Why does the procedure of Chapter 3 not work for step 3 of the reduction process above? (This is elimination of higher terms.)

(8) Which combinations of singularities would you expect to occur at various levels in a versal unfolding of $t^{k+1}$ (at $t = 0$), close to $t^{k+1}$ itself? The pictures at the beginning of the chapter suggest for $k = 4$: $A_4$ or $A_1/A_3$ or $A_2/A_2$ or $A_1/A_1/A_2$ or $A_1 A_2/A_1$ (with the $A_2$ and the 'first' $A_1$ at the *same* level) or $A_1/A_1/A_1/A_1$ or $A_1 A_1/A_1/A_1$ or $A_1 A_1/A_1 A_1$ (two $A_1$s at one level and two at another), but *not* $A_2 A_2$ (two $A_2$s at the same level) or $A_1 A_3$ or $A_1 A_1/A_2$. Try $k = 5$. It is really a matter of picturing all possible polynomials $t^{k+1} +$ lower terms, where the lower terms have small coefficients.

Now we turn to two matters: how to recognize a (p)versal unfolding when you see one, and what to do with it once you have recognized it. For us the crucial property we want to exploit concerns the bifurcation set of $F$, which has appeared before in chapter 1.

As a practical suggestion, we suggest that the statements of 6.9p, 6.10p, 6.13p and 6.14p below are understood, and the proofs postponed. The important thing is to see the good geometrical consequences of these high-faluting considerations, and those start in chapter 7.

### 6.9p     Criterion for (p)versality

Let $F$, as in 6.1, be an unfolding of $f$, which has type $A_k$ $(k \geqslant 1)$ at $t_0$. Then $F$ is (p)versal if and only if every real polynomial $p(t)$ of degree $\leqslant k-1$ and without constant term can be written in the form

$$p(t) = \sum_{i=1}^{r} c_i j^{k-1} \left( \frac{\partial F}{\partial x_i} (t, x_0) \right) (t_0)$$

for real constants $c_i$, where $j^{k-1}$ denotes the $(k-1)$-jet.

There are two equivalent formulations of this criterion.

### 6.10p

Let $j^{k-1}(\partial F/\partial x_i(t, x_0))(t_0) = \alpha_{1i}t + \alpha_{2i}t^2 + \cdots + \alpha_{k-1,i}t^{k-1}$ for $i = 1, \ldots, r$. Then $F$ is (p)versal iff the $(k-1) \times r$ matrix of coefficients $\alpha$ has rank $k-1$. (This certainly requires $k-1 \leqslant r$, so the smallest possible value of $r$ is $k-1$.)

**6.11p**     This is for ring enthusiasts. Let $\mathbb{R}[t]$ denote the ring of polynomials in $t$ and let $\mathfrak{m}$ denote the maximal ideal consisting of polynomials with zero constant term. Finally let $\langle t^k \rangle$ denote the ideal of polynomial multiples of $t^k$. Then $F$ is (p)versal iff the above $(k-1)$-jets span the real vector space $\mathfrak{m}/\langle t^k \rangle$.

To work out the $(k-1)$-jet, you find $\partial F/\partial x_i(t, x_0)$, which is a function of $t$, and then expand about $t = t_0$, or work out the successive derivatives of this function at $t_0$ for the Taylor expansion. For example, take $F$ to be the unfolding $G$ of 6.6p. Then $\partial G/\partial x_i = t^i$ (independently of $x$), so the $(k-1)$-jet of $\partial G/\partial x_i(t, 0)$ is the $(k-1)$-jet of $t^i$ at 0, i.e. $t^i$ since $i \leqslant k-1$. Since every polynomial $p(t)$ as in 6.9p can certainly be written as a linear combination of $t, t^2, \ldots, t^{k-1}$, $G$ does satisfy the criterion. The matrix of 6.10p is just the identity $(k-1) \times (k-1)$ matrix in this case – hence rank $k-1$.

It is worth noting that if $r \geqslant k-1$ then almost any choice of coefficients $\alpha_{ji}$ as in 6.10p will make the rank $k-1$ and so give a (p)versal unfolding.

For the rank will drop below $k-1$ only when all $(k-1) \times (k-1)$ minors of the matrix are zero. This says that, *provided $r \geqslant k-1$, almost every unfolding is (p)versal.*

### 6.12p  Exercises

(1) Show that the criteria in 6.9p and 6.10p are equivalent. (This is just linear algebra, for the set of $p(t)$ is a real vector space of dimension $k-1$.)

(2) Ring enthusiasts should verify the equivalence of 6.9p and 6.11p. (Once you know the definitions this is virtually immediate.)

(3) Here are some unfoldings, as in 6.1, where $t_0 = 0$ and $x_0 = 0$. In each case determine (i) the type $A_k$ of $f = F_0$; (ii) the $(k-1)$-jets of $\partial F/\partial x_i(t, 0), i = 1, \ldots, r$; (iii) the matrix of 6.10p; (iv) whether $F$ is a (p)versal unfolding of $f$ at 0.

  (a) $r = 3$, $F(t, x) = t^4 + x_1 t^3 + x_2 t^2 + x_3 t$,

  (b) $r = 2$, $F(t, x) = t^4 + x_1 x_2 t^2 + x_2 t$,

  (c) $r = 3$, $F(t, x) = t^5 + x_2^2 t^3 + x_1 t^3 + x_2 t^2 + x_3(t+1)t$.

(4) (Harder.) Show that, for any integer $l \geqslant 1$, the unfolding $F(t, x_1, \ldots, x_{k-1})$ $= t^{k+1} + x_1(t+1)^l + x_2(t+1)^{l+1} + \cdots + x_{k-1}(t+1)^{l+k-2}$ is a (p)versal unfolding of $F_0$ at 0.

***Proof of 6.9p*** Assume that the jet condition holds. We take $f(t) = t^{k+1}$ and leave the minor modifications needed for any function of type $A_k$ to the reader. Let $G$ be as in 6.6p: then $G$ *is* (p)versal so we can write

$$F(t, x) = G(a(t, x), b(x)) + c(x)$$

as in 6.3p. Hence

$$\frac{\partial F}{\partial x_i}(t, x_0) = \frac{\partial G}{\partial t}(t, 0)\frac{\partial a}{\partial x_i}(t, x_0) + \sum_{l=1}^{k-1} \frac{\partial G}{\partial u_l}(t, 0)\frac{\partial b_l}{\partial x_i}(x_0) + \frac{\partial c}{\partial x_i}(x_0)$$

$$= (k+1)t^k \frac{\partial a}{\partial x_i}(t, x_0) + \sum_{l=1}^{k-1} t^l \frac{\partial b_l}{\partial x_i}(x_0) + \text{constant}.$$

Now the first and last terms on the right do not contribute to the $(k-1)$-jet so the matrix of $(\partial b_l/\partial x_i(x_0))$ has rank $k-1$, by 6.10p. We use this fact to prove that $F$ is versal.

Let $F_1: \mathbb{R} \times \mathbb{R}^s, (0, w_0) \to \mathbb{R}$ be any unfolding of $f$, where we use coordinates $w_1, \ldots, w_s$ in $\mathbb{R}^s$. Then, as in 6.2p, we can find functions $A(t, w)$, $B(w)$, $C(w)$ with

$$G(t, B(w)) = F_1(A(t, w), w) + C(w), \quad A(t, w_0) = t$$

and functions $\alpha(t, x)$, $b(x)$, $c(x)$ with

$$G(t, b(x)) = F(\alpha(t, x), x) + c(x), \quad \alpha(t, x_0) = t.$$

From these we try to induce $F_1$ from $F$.

Consider the $k-1$ equations $b(x) - B(w) = 0$ in variables $x_1, \ldots, x_r$, $w_1, \ldots, w_s$. Since $(\partial b_l/\partial x_i(x_0))$ has rank $k-1$ the Implicit Function Theorem implies that we can solve these equations by expressing the $x$s as functions of the $w$s: say $b(B_1(w)) - B(w) = 0$ for all $w$ close to $w_0$, where $B_1(w_0) = x_0$. Thus

$$F_1(A(t, w), w) + C(w) = G(t, B(w)) = F(\alpha(t, B_1(w)), B_1(w)) + c(B_1(w)).$$

Also $\partial A/\partial t(0, w_0) = 1 \neq 0$ so again by the Implicit Function Theorem we can find $A_1(t, w)$ with $A(A_1(t, w), w) = t$ for $(t, w)$ close to $(0, w_0)$. In fact $A_1 \times id = (A \times id)^{-1}$ where $id: \mathbb{R}^s \to \mathbb{R}^s$ is the identity map, so that $A_1(t, w_0) = t$. Hence

$$F_1(t, w) = F(\alpha(A_1(t, w), B_1(w)), B_1(w)) + c(B_1(w)) - C(w)$$

and since $\alpha(A_1(t, w_0), B_1(w_0)) = \alpha(t, x_0) = t$ we have (p)induced $F_1$ from $F$ and the versality of $F$ follows.

The converse implication is much easier and is left as an exercise. (Given $F$ is (p)versal, the unfolding $G$ of 6.6p can be (p)induced from it.)     $\square$

### 6.13p     Definitions

Let $F$ be an unfolding of $f$, as in 6.1. The *singular set* $S_F$ of $F$ consists of pairs $(t, x)$ for which $F_x$ is singular at $t$:

$$S_F = \{(t, x) \in \mathbb{R} \times \mathbb{R}^r : \partial F/\partial t = 0 \text{ at } (t, x)\}.$$

The *bifurcation set* of $F$ is the set

$$\mathscr{B}_F = \{x \in \mathbb{R}^r : \text{there exists } t \text{ with } \partial F/\partial t = \partial^2 F/\partial t^2 = 0 \text{ at } (t, x)\}.$$

If $x \in \mathscr{B}_F$ and $\partial F/\partial t = \partial^2 F/\partial t^2 = 0$ at $(t, x)$, then $t$ is said to *correspond* to $x$.

If we suppose that $F$ is a (p)*versal* unfolding of $f = F_{x_0}$ at $t_0$, and if $(t_0, x_0) \in S_F$, then $(t_0, x_0)$ is a regular point of the function $\mathbb{R} \times \mathbb{R}^r \to \mathbb{R}$ given by $(t, x) \mapsto \partial F/\partial t$. (For $f$ has a singularity $A_{\geqslant 1}$ at $t_0$ and (p)versality implies that the 1-jets of $\partial F/\partial x_i(t, x_0)$ cannot all be zero, so not all $\partial^2 F/\partial t \partial x_i(t_0, x_0)$ can be zero.) Hence $S_F$ is a parametrized $r$-manifold in $\mathbb{R}^{r+1}$ in a neighbourhood $U_1$ of $(t_0, x_0)$. Writing $\pi: S_F \to \mathbb{R}^r$ for the projection $\pi(t, x) = x$, $\mathscr{B}_F$ is locally the set of critical values of this projection: this is exactly analogous to 5.21, $\partial F/\partial t$ and $\mathscr{B}_F$ here playing the roles of $F$ and $\mathscr{D}_F$ there.

Now let $F$ and $G$ be any two (p)versal $r$-parameter unfoldings of $f$ (at $t_0$) and $g$ (at $t_1$) respectively, both of type $A_k$ ($k \geqslant 1$). Thus

$$F: \mathbb{R} \times \mathbb{R}^r, (t_0, x_0) \to \mathbb{R}$$

$$G: \mathbb{R} \times \mathbb{R}^r, (t_1, u_0) \to \mathbb{R}$$

where $F(t, x_0) = f(t)$ ($t$ close to $t_0$), $G(t, u_0) = g(t)$ ($t$ close to $t_1$), $f$ having type $A_k$ at $t_0$ and $g$ having type $A_k$ at $t_1$. Then by the above remarks there exist neighbourhoods $U$ of $(t_0, x_0)$ and $V$ of $(t_1, u_0)$ in $\mathbb{R} \times \mathbb{R}^r$ such that $S_F \cap U$ and $S_G \cap V$ are parametrized $r$-manifolds.

### 6.14p Proposition

*With the above notation, the neighbourhoods $U$ and $V$ can be chosen so that there exists a diffeomorphism $\phi: U \to V$ of the form $\phi(t, x)$ $=(a(t, x), b(x))$ where $a(t_0, x_0)=t_1$, $b(x_0)=u_0$ and*

*(i) $\phi(S_F \cap U)=S_G \cap V$,*

*(ii) $b$ is a diffeomorphism from $\pi(U)$ to $\pi(V)$, where each $\pi$ is projection to the unfolding parameters, and $b(\mathscr{B}_F \cap \pi(U))=\mathscr{B}_G \cap \pi(V)$.*

Thus the bifurcation sets $\mathscr{B}_F$ and $\mathscr{B}_G$ are locally diffeomorphic (fig. 6.2).

It follows that we can meaningfully speak of *the* bifurcation set of an $r$-parameter (p)versal unfolding of an $A_k$ singularity: the local picture is the same, up to diffeomorphism, for all.

***Proof*** Now $f$ and $g$ are $\mathscr{R}$-equivalent, say $f(t)=g(h(t))+c$ ($h$ a diffeomorphism $\mathbb{R}$, $t_0 \to \mathbb{R}$, $t_1$). Let $G_1(t, x)=G(h(t), x)+c$. Then replacing $F$ by $G_1$ the proposition is very easy to prove: take $\phi(t, x)=(h(t), x)$, so that $b(x)=x$. (Here $U$ is chosen small enough to make $\phi$ a diffeomorphism onto $\phi(U)$.) We prove below that the proposition is true with $G$ replaced by $G_1$ (this amounts to the simplification $t_0=t_1$, $f=g$); composing the two $\phi$s then proves the required result.

So now assume $t_0=t_1$, $f=g$. We also take $r$ to be the smallest possible number, namely $k-1$: the result for general $r$ is deduced in exercise 6.15p(2) below. Since $G$ is (p)versal, there exist $a, b, c$ as in 6.5p with

$$F(t, x)=G(a(t, x), b(x))+c(x)$$

**Fig. 6.2.** Diffeomorphic bifurcation sets

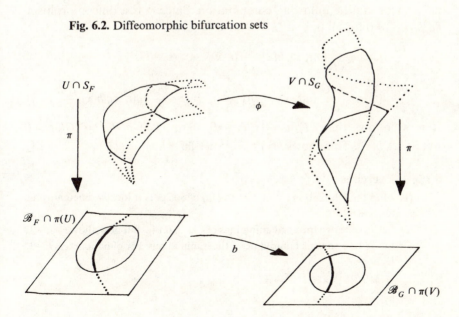

and $a(t, x_0) = t$ for $t$ close to $t_0$. We prove that $\phi = (a, b)$ and $b$ are the local diffeomorphisms required. First consider $b$.

We have

$$\frac{\partial F}{\partial x_i} = \frac{\partial G}{\partial t}\frac{\partial a}{\partial x_i} + \sum_{l=1}^{k-1} \frac{\partial G}{\partial u_l}\frac{\partial b_l}{\partial x_i} + \frac{\partial c}{\partial x_i}$$

where we write $G$ as a function of $t, u_1, \ldots, u_{k-1}$. Since $G(t, u_0)$ has type $A_k$ at $t_0$, $\partial G/\partial t(t, u_0)$ has type $A_{k-1}$ at $t_0$, so putting $x = x_0$, $u = u_0$ and taking $(k-1)$-jets of the above formula gives

$$j^{k-1}\left(\frac{\partial F}{\partial x_i}(t, x_0)\right)(t_0) = \sum_{l=1}^{k-1} j^{k-1}\left(\frac{\partial G}{\partial u_l}(t, u_0)\right)(t_0)\frac{\partial b_l}{\partial x_i}(x_0).$$

The matrices formed by coefficients in the $(k-1)$-jets of the $\partial F/\partial x_i(t, x_0)$ and $\partial G/\partial u_l(t, u_0)$ are both nonsingular, by the (p)versality criterion 6.10p. Calling these matrices $P$ and $Q$ the above equation says

$$P = Q\left(\frac{\partial b_l}{\partial x_i}(x_0)\right).$$

Hence the Jacobian matrix of $b$ at $x_0$ is nonsingular and $b$ is a local diffeomorphism.

By hypothesis $\partial a/\partial t(t_0, x_0) = 1$ and it follows that the Jacobian matrix of $\phi$ at $(t_0, x_0)$ is also nonsingular, so that $\phi$ is also a local diffeomorphism, and shrinking $U$ if necessary we have proved (i).

It remains to check that $b$ takes $\mathscr{B}_F$ to $\mathscr{B}_G$. One can either do this by proving that $b$ must take the set of critical values of $\pi|S_F \cap U$ to those of $\pi|S_G \cap V$ or via the following computation. Since $\phi$ is a diffeomorphism, $\partial a/\partial t(t, x) \neq 0$ for all $(t, x) \in U$. We have

$$\frac{\partial F}{\partial t}(t, x) = \frac{\partial G}{\partial t}(a(t, x), b(x))\frac{\partial a}{\partial t}(t, x),$$

$$\frac{\partial^2 F}{\partial t^2}(t, x) = \frac{\partial^2 G}{\partial t^2}(a(t, x), b(x))\left(\frac{\partial a}{\partial t}(t, x)\right)^2 + \frac{\partial G}{\partial t}(a(t, x), b(x))\frac{\partial^2 a}{\partial t^2}(t, x).$$

It now follows that $\partial F/\partial t = \partial^2 F/\partial t^2 = 0$ at $(t, x)$ iff $\partial G/\partial t = \partial^2 G/\partial t^2 = 0$ at $(a(t, x), b(x))$. This shows $b(x) \in \mathscr{B}_G \cap \pi(V)$ iff $x \in \mathscr{B}_F \cap \pi(U)$ □

### 6.15p    Exercises

(1) Find the bifurcation set of $F$ as in (3a) of 6.12p. Is it locally diffeomorphic to a cusp?

(2) Let $F, G$ be two (p)versal unfoldings of $f$ as in 6.14p, but now allow $r \geqslant k-1$. Let $H$ be a (p)versal unfolding of the minimal number of parameters $k-1$. Define

$$\tilde{H}: \mathbb{R} \times \mathbb{R}^{k-1} \times \mathbb{R}^{r-k+1} \to \mathbb{R}$$

by

$$\tilde{H}(t, y, z) = H(t, y).$$

Show that $S_{\tilde{H}} = S_H \times \mathbb{R}^{r-k+1}$ and $\mathcal{B}_{\tilde{H}} = \mathcal{B}_H \times \mathbb{R}^{r-k+1}$. Since $H$ is versal we have $F(t, x) = H(a(t, x), b(x)) + c(x)$ as usual. Show that the Jacobian matrix of $b$ at $x_0$ has rank $k-1$. (This is similar to the first part of the proof of 6.9p. What is the connexion between this Jacobian matrix and the 'jet matrices' as in 6.10p for $F$ and $H$?) Now show that linear maps $l_i: \mathbb{R}^r \to \mathbb{R}$, $i = 1, \ldots, r-k+1$, can be chosen so that the maps $\tilde{b} = (b, l_1, \ldots, l_{r-k+1})$: $\mathbb{R}^r \to \mathbb{R}^r$, $(a, \tilde{b}): \mathbb{R} \times \mathbb{R}^r \to \mathbb{R} \times \mathbb{R}^r$ are local diffeomorphisms at $x_0$ and $(t_0, x_0)$ respectively. Show that $(a, \tilde{b})$ maps $S_F$ locally to $S_{\tilde{H}}$ and $b$ takes $\mathcal{B}_F$ locally to $\mathcal{B}_{\tilde{H}}$, as in the proof of 6.14p, and deduce the general result of 6.14p.

We shall now look at these bifurcation sets for small values of $r$ and $k$. Of course, $r \geqslant k-1$.

### 6.16p    (p) Versal unfoldings of $A_2$ ($r \geqslant 1$)

For $r = 1$ we need only consider $\mathcal{B}_F$ where

$$F: \mathbb{R} \times \mathbb{R}, (0, 0) \to \mathbb{R}$$

is given by

$$F(t, x) = t^3 + xt.$$

The bifurcation set $\mathcal{B}_F$ is $\{x \in \mathbb{R}: \text{for some } t, 3t^2 + x = 6t = 0\}$, which is just $\{0\}$, i.e. a point of $\mathbb{R}$.

For higher values of $r$ we just take

$$\tilde{F}: \mathbb{R} \times \mathbb{R}^r, (0, 0) \to \mathbb{R}$$

given by

$$\tilde{F}(t, x) = t^3 + x_1 t \text{ (independent of } x_2, \ldots, x_r).$$

Then $\tilde{\mathcal{B}}_F = \mathcal{B}_F \times \mathbb{R}^{r-1} = \{0\} \times \mathbb{R}^{r-1}$. Hence the local picture for the bifurcation set of an $r$-parameter (p)versal unfolding of an $A_2$ singularity is a linear space of dimension $r-1$ in $\mathbb{R}^r$. That is, the bifurcation set of any (p)versal $r$-parameter unfolding of an $A_2$ singularity is locally a parametrized $(r-1)$-manifold in $\mathbb{R}^r$ (fig. 6.3).

**Fig. 6.3.** The fold

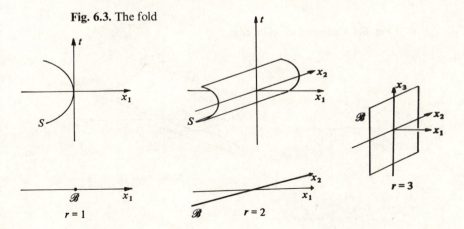

In the first two diagrams the set $\partial F/\partial t = 0$ in $\mathbb{R} \times \mathbb{R}^r$ is also drawn: the projection to the bifurcation set is called a *fold*.

**6.17p    (p)Versal unfoldings of $A_3$ $(r \geqslant 2)$**

Let

$$F: \mathbb{R} \times \mathbb{R}^r, (0, 0) \to \mathbb{R} \quad (r \geqslant 2)$$

be

$$F(t, x) = \tfrac{1}{4}t^4 + \tfrac{1}{2}x_1 t^2 + x_2 t.$$

Then $F$ is a (p)versal unfolding of $f(t) = \tfrac{1}{4}t^4$ (at $t_0 = 0$) and

$$\mathscr{B}_F = \{x: t^3 + x_1 t + x_2 = 3t^2 + x_1 = 0 \text{ for some } t\}$$
$$= \{x: x_1 = -3t^2, x_2 = 2t^3 \text{ for some } t\}$$
$$= \{x: 4x_1^3 + 27x_2^2 = 0\}.$$

For $r = 2$ this is the now familiar cusp, also met as the discriminant of a cubic polynomial; for $r = 3$ it is a cuspidal edge (fig. 6.4). Of course a suitable local diffeomorphism of $\mathbb{R}^2$ will take the cusp to $\{x: x_1^3 = x_2^2\}$. See fig. 1.6 for a picture of the surface $\partial F/\partial t = 0$ and the projection to $\mathbb{R}^2$ in the case $r = 2$.

**6.18p    (p)Versal unfoldings of $A_4$ $(r \geqslant 3)$**

Let

$$F: \mathbb{R} \times \mathbb{R}^r, (0, 0) \to \mathbb{R} \quad (r \geqslant 3)$$

be

$$F(t, x) = \tfrac{1}{5}t^5 + \tfrac{1}{3}x_1 t^3 + \tfrac{1}{2}x_2 t^2 + x_3 t.$$

Then $F$ is a (p)versal unfolding of $f(t) = \tfrac{1}{5}t^5$ (at $t_0 = 0$) and

$$\mathscr{B}_F = \{x: t^4 + x_1 t^2 + x_2 t + x_3 = 4t^3 + 2x_1 t + x_2 = 0 \text{ for some } t\}.$$

For $r = 3$ this is exactly the discriminant of a quartic polynomial, as in 5.36, i.e. the swallowtail. See fig. 5.18.

**Fig. 6.4.** Cusp and cuspidal edge

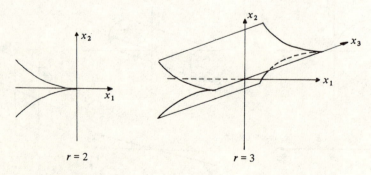

$r = 2$                                    $r = 3$

Thus the bifurcation set of any $r$-parameter versal unfolding of an $A_k$ singularity, with $3 \geqslant r \geqslant k-1$, is locally diffeomorphic to a neighbourhood of the origin in one of the pictures (including the swallowtail). In the next chapter many specific applications of this fact will be given, and you could now turn to the applications in 7.1 to 7.6.

There are two further matters we consider in this chapter: first the different kinds of cusp, and second the connexion between bifurcation sets and discriminants, where we restate the results above in the 'non-p' case.

## Cusps

Is it possible that the two pictures – line and cusp – for $r=2$ are actually the same? That is, could a local diffeomorphism of $\mathbb{R}^2$ take one onto the other? It is not hard to show that this cannot happen: there is no diffeomorphism $\phi: \mathbb{R}^2, 0 \to \mathbb{R}^2, 0$ taking the $x_2$-axis near 0 to the curve $x_1^3 = x_2^2$. (What is wrong with $\phi(x_1, x_2) = (x_1 - x_2^2, x_2^3)$?) This and other things are the subject of the following exercises.

### 6.19 Exercises

(1) Suppose $\phi$ as above exists, write $\phi_1, \phi_2$ for the components of $\phi$ and let $f(y) = \phi_2(0, y)$, $g(y) = \phi_1(0, y)$ so that $f^2 = g^3$ for all $y$ close to 0. Use Taylor's theorem (see 3.8(1)) to show that, for all small $y$,

$f(y) = a_1 y + a_2 y^2 + y^3 f_1(y)$,

$g(y) = b_1 y + y^2 g_1(y)$

for constants $a_1$, $a_2$, $b_1$ and smooth $f_1$ and $g_1$. Use $f^2 = g^3$ to show that $a_1 = b_1 = 0$ and deduce that in fact $\phi$ fails to be a local diffeomorphism at 0.

(2) The cusp $x_1^3 = x_2^2$ is called an *ordinary cusp*; so is anything obtained from it by a local diffeomorphism of $\mathbb{R}^2$ at 0. How about extraordinary cusps? The set $\{x \in \mathbb{R}^2 : x_1^5 = x_2^2\}$ is the least extraordinary and is called a *rhamphoid cusp*. Show by the following steps that there is no local diffeomorphism of $\mathbb{R}^2$ taking an ordinary to a rhamphoid cusp. Suppose that $\phi: \mathbb{R}^2, 0 \to \mathbb{R}^2, 0$ has $(\phi_1(t^2, t^3))^5 = (\phi_2(t^2, t^3))^2$ for all small $t$. Write $f(t) = \phi_1(t^2, t^3)$, $g(t) = \phi_2(t^2, t^3)$.

(i) Show $f(0) = f'(0) = 0$ and deduce that the 9-jet of $f^5$ at 0 is zero.

(ii) Show $\frac{1}{2} g''(0) = \partial \phi_2/\partial x_1(0, 0)$; $\frac{1}{6} g'''(0) = \partial \phi_2/\partial x_2(0, 0)$. Deduce that $g(t) = At^2 + Bt^3 + Ct^4 + t^5 g_1(t)$ where $A = \partial \phi_2/\partial x_1(0, 0)$, $B = \partial \phi_2/\partial x_2(0, 0)$, $C \in \mathbb{R}$ and $g_1$ is smooth.

(iii) Putting $f^5 = g^2$ show $A = B = 0$. Why does this give a contradiction?

(iv) What is wrong with $\phi(x_1, x_2) = (x_1^3, x_2^5)$?

(3) Show that there is no local diffeomorphism $\phi: \mathbb{R}^2, 0 \to \mathbb{R}^2, 0$ taking the $x_1$-axis to itself and the $x_2$-axis to the curve $x_2 = x_1^3$. (Thus you want to show $\phi_2(x_1, 0) = 0$ and $(\phi_1(0, x_2))^3 = \phi_2(0, x_2)$ is impossible.) Note that the two axes cross without touching but the $x_1$-axis and the cubic curve touch as they cross.

## Versal unfoldings

To a family $F: \mathbb{R} \times \mathbb{R}^r, (t_0, x_0) \to \mathbb{R}$ there is associated another family $\partial F/\partial t$ with the same domain. Clearly the bifurcation set of $F$ coincides with the discriminant set, as defined in 5.3, of $\partial F/\partial t$. Furthermore if the function $F(-, x_0)$ has type $A_k$ $(k \geqslant 1)$ at $t_0$, then $\partial F/\partial t(-, x_0)$ has type $A_{k-1}$ at $t_0$.

In chapter 5 we met discriminant sets as envelopes of families and many of the families did not arise, at any rate in an obvious way, as derivatives $\partial F/\partial t$ of other families $F$. There is a local structure theorem for discriminants just as there is for bifurcation sets, and we should like to apply it to all the families of chapter 5. A slightly different concept of 'versal unfolding' is appropriate here, and of course the main thing is to be able to recognize and use these versal unfoldings as they arise. Rather than go through the whole theory again we summarize here the new forms of definitions and results. The changes to proofs are extremely slight.

We are still talking about $r$-parameter unfoldings of $f$ of the form

$$F: \mathbb{R} \times \mathbb{R}^r, (t_0, x_0) \to \mathbb{R}$$

where $f(t) = F(t, x_0)$.

**6.3**    Let $G: \mathbb{R} \times \mathbb{R}^k \to \mathbb{R}$ be

$$G(t, u) = u_1 + u_2 t + u_3 t^2 + \cdots + u_k t^{k-1} + t^{k+1}$$

where $u = (u_1, \ldots, u_k)$. Then any unfolding $\bar{F}$ of $t^{k+1}$ can be written

$$\bar{F}(t, x) = G(a(t, x), b(x))$$

where $a(t, 0) = t$ for all $t$ close to 0. (Here $t_0 = 0$.)

**6.5**    **Induced and versal unfoldings**
This is as in 6.5p, but $c$ is identically zero. Thus when $F(t, x) = G(a(t, x), b(x))$, where $a(t, x_0) = t$, for all $t$ close to $t_0$, we say that $F$ is induced from $G$. Notice that, for a given $x$, $F_x$ at $t$ is still $\mathcal{R}$-equivalent to $G_{b(x)}$ at $a(t, x)$, provided $(t, x)$ is sufficiently close to $(t_0, x_0)$. The $\mathcal{R}$-equivalence is of the special kind where no additive constant is used.

**6.6**    *The unfolding $G: \mathbb{R} \times \mathbb{R}^k \to \mathbb{R}$, given by*

$$G(t, x) = \pm t^{k+1} + x_1 + x_2 t + \cdots + x_k t^{k-1}$$

is a versal unfolding of $g(t) = \pm t^{k+1}$ at $t_0 = 0$. It is miniversal, having the minimum number of parameters, $k$, for an $A_k$ singularity.

### 6.9 Criterion for versality

*Let F be as above and let f have type $A_k$ $(k \geqslant 1)$. Then F is versal if and only if every real polynomial $p(t)$ of degree $\leqslant k-1$ can be written as a real linear combination of the $(k-1)$-jets with constant of $\partial F/\partial x_i(t, x_0)$ $(i = 1, \ldots, r)$ at $t_0$:*

$$p(t) = \sum_{i=1}^{r} c_i \left( \frac{\partial F}{\partial x_i}(t_0, x_0) \right) + j^{k-1} \left( \frac{\partial F}{\partial x_i}(t, x_0) \right)(t_0)$$

*where the $c_i$ are real numbers.*

### 6.10 Matrix criterion for versality

*Write the $(k-1)$-jet with constant as $\alpha_{0i} + \alpha_{1i}t + \alpha_{2i}t^2 + \cdots + \alpha_{k-1,i}t^{k-1}$.*
*Then F is versal iff the $k \times r$ matrix of coefficients $\alpha$ has rank $k$ (of course this is possible only if $r \geqslant k$).*

### 6.11 Ring criterion for versality

*The $(k-1)$-jets with constant have to span the real vector space* $\mathbb{R}[t]/\langle t^k \rangle$.

### 6.14 Uniqueness of the discriminant set

*With the hypotheses of 6.14p, omitting the affix $(p)$, the discriminant sets $\mathcal{D}_F$ and $\mathcal{D}_G$ are locally diffeomorphic.* For the precise meaning of this, we replace $S$ by $M$ in 6.14p, where

$$M_F = \{(t, x) \in \mathbb{R} \times \mathbb{R}^r : F(t, x) = 0\},$$

$$\mathcal{D}_F = \{x \in \mathbb{R}^r : \text{there exists } t \in \mathbb{R} \text{ with } F = \partial F/\partial t = 0 \text{ at } (t, x)\}.$$

The proof proceeds as before, by showing that $\phi = (a, b)$ and $b$ are local diffeomorphisms.

### 6.16 Versal unfoldings of $A_1$ $(r \geqslant 1)$
Take

$$F(t, x) = t^2 + x_1 \quad x = (x_1, \ldots, x_r).$$

The discriminant sets are the same pictures as in 6.16p, namely each is a linear subspace of $\mathbb{R}^r$. Thus the fold is here associated with $A_1$ singularities.

### 6.17 Versal unfoldings of $A_2$ $(r \geqslant 2)$
Take

$$F(t, x) = t^3 + x_1 t + x_2 \quad x = (x_1, \ldots, x_r).$$

The discriminant sets are as in 6.17p (cusp, cuspidal edge).

### 6.18    Versal unfoldings of $A_3$ $(r \geqslant 3)$

Take

$$F(t, x) = t^4 + x_1 t^2 + x_2 t + x_3.$$

The discriminant set for $r = 3$ is the swallowtail.

Applications of these results will be found in the next chapter, starting at 7.7.

### 6.20    Exercises

All the proofs and exercises in the 'potential' situation can be adapted to the present case by minor modifications.

(1) If you have read the proofs of 6.9p and 6.14p then you should find the minor changes needed to prove 6.9 and 6.14.

(2) Adapt example 6.7p by taking $G(t, x_1, x_2) = t^3 + x_1 + x_2 t$ and $H(t, x_1, x_2, x_3)$ $= t^3 + x_1 + x_2 t + x_3 t^2$. Again each can be induced from the other.

(3) In exercise 6.8p(1) just delete $c(x)$ and $C(y)$, together with all affixes (p).

(4) From (2) above you should be able to see how to adapt exercise 6.8p(2). Exercises (4) and (5) of 6.8p do not require any change.

(5) One way to adapt exercise 6.12p(3) is to differentiate the given $F(t, x)$ with respect to $t$ and take these as the new unfoldings. In fact, given a 'potential function' $f$, of type $A_k$ at $t_0$, then verifying that $F$ is a (p)versal unfolding of $f$ at $t_0$ is the same as verifying that $\partial F/\partial t$ is a versal unfolding of $f'$ at $t_0$.

### 6.21    Perturbing an unfolding

After 4.18 it is shown that any smooth function $f: I \to \mathbb{R}$ ($I$ an open interval in $\mathbb{R}$) can be 'perturbed' into a function $f(t) + u_1 t + u_2 t^2$ which has *only* $A_1$ singularities. Here $u = (u_1, u_2)$ can be any point outside a certain null set in $\mathbb{R}^2$, and in particular can be arbitrarily close to $(0, 0)$. A similar argument shows that any *family* $F: I \times \mathbb{R}^r \to \mathbb{R}$ of functions $F_x$ $(x \in \mathbb{R}^r)$ can be perturbed into $\Phi(t, x, u) = F(t, x) + u_1 t + \cdots + u_{r+2} t^{r+2}$ where, for each fixed $u \in \mathbb{R}^{r+2}$ outside a null set, and all $x$, this function of $t$ has only $A_{\leqslant r+1}$ singularities. (This can be done directly, as in the example following 4.18, or using Thom's Transversality Lemma – see 8.20(5).) In particular $u$ can be chosen arbitrarily close to $0 \in \mathbb{R}^{r+2}$ and $F$ is then *arbitrarily close to a family* $\Phi_u$ containing only $A_{\leqslant r+1}$ singularities.

Let us put the matter less precisely. Given a family $F$ as above, consider

$$F^k: I \times \mathbb{R}^r \to \mathbb{R}^k$$

where

$$(t, x) \mapsto \left( \frac{\partial F}{\partial t}, \ldots, \frac{\partial^k F}{\partial t^k} \right).$$

Now when $k > r+1$ the regular values of $F^k$ are precisely the points of $\mathbb{R}^k$ *not* in the image of $F^k$. Sard's Theorem says that *almost all* points of $\mathbb{R}^k$

are regular values, and in particular we would *expect* 0 to be a regular value, that is we would *expect* there to be no points $(t, x)$ for which $\partial F/\partial t = \cdots = \partial^k F/\partial t^k = 0$. This suggests that 'in general', no function $F_x$ has an $A_{>r+1}$ singularity. Even less precisely, we do not expect $r+1$ variables $(t, x)$ to be able to satisfy $k > r+1$ equations $\partial F/\partial t = \cdots = \partial^k F/\partial t^k = 0$.

When $k \leq r+1$ the condition for 0 to be a regular value of $F^k$ is easily computed. Suppose $F^k(t, x) = 0$. The Jacobian matrix of $F^k$ at $(t, x)$ then has the form

$$\begin{pmatrix} 0 & \dfrac{\partial^2 F}{\partial t \partial x_1} & \cdots & \dfrac{\partial^2 F}{\partial t \partial x_r} \\[2mm] \vdots & \vdots & & \vdots \\[2mm] 0 & \dfrac{\partial^k F}{\partial t^{k-1} \partial x_1} & \cdots & \dfrac{\partial^k F}{\partial t^{k-1} \partial x_r} \\[2mm] \dfrac{\partial^{k+1} F}{\partial t^{k+1}} & \dfrac{\partial^{k+1} F}{\partial t^k x_1} & \cdots & \dfrac{\partial^{k+1} F}{\partial t^k x_r} \end{pmatrix}$$

Now if this matrix has rank $k$ and $\partial^{k+1} F/\partial t^{k+1}(t, x) \neq 0$ (i.e. if $F_x$ has type $A_k$ at $t$) then the indicated block of the matrix has rank $k-1$. However multiplying the $i$th row by $1/i!$ (this does not change the rank) the latter condition is precisely the condition that $F$ is a (p)versal unfolding of $F_x$ at $t$. Thus if 0 is a regular value of $F^k$ then whenever $F_x$ has type $A_k$ at $t$, $F$ is a (p)versal unfolding of $F_x$ at $t$. Since Sard's Theorem implies that we can *expect* 0 to be a regular value of $F^k$, this suggests that 'in general' $F$ will (p)versally unfold all singularities of type $A_k$, $k \leq r+1$.

Again we can perturb $F$ by a suitable $\Phi$ and verify that *$F$ is arbitrarily close to families which contain only $A_{\leq r+1}$ singularities, all of which are (p)-versally unfolded by the perturbed family $\Phi_u$.* For this it is best to use Thom's Transversality Lemma – see 8.20(6).

In a similar way one can verify that $F$ is arbitrarily close to families which contain only $A_{\leq r}$ singularities at points where the functions in question take the value 0, and that all these singularities are versally unfolded by the perturbed family.

## 6.22 Project

In this chapter we have described the bifurcation sets of (p)versal unfoldings of $A_k$ singularities, $2 \leq k \leq 4$ (and the discriminant sets of versal unfoldings of $A_k$ singularities, $1 \leq k \leq 3$). The next bifurcation set to consider is that associated with an $A_5$ singularity. The problem with this set is that it lies in a 4-dimensional space (as does the discriminant set of an $A_4$). Investigate the geometry of this so-called *butterfly* set using the pictures and discussion in Poston and Stewart (1978), pp 178–180, Bröcker and Lander (1975), pp 162–5, and Poston and Woodcock (1974). (The butterfly set does arise naturally when discussing for example the geometry of a space curve that is moved in $\mathbb{R}^3$ by a generic 'isotopy' or family of self-diffeomorphisms of $\mathbb{R}^3$.)

# 7

## *Unfoldings: applications*

'Singularity is almost invariably a clue.'
(*The Boscombe Valley Mystery*)

In this chapter we apply the rather technical ideas of chapter 6 to some very concrete geometrical problems. Although the ideas are technical (and their proofs even more so), applying them is, fortunately, relatively easy. We shall concentrate on elucidating the local structure of bifurcation and discriminant sets (such as envelopes), though this is by no means the only application.

Here is the pattern. We start with a family $F: \mathbb{R} \times \mathbb{R}^r \rightarrowtail \mathbb{R}$, where $r = 2$ or 3, whose bifurcation or discriminant set interests us. If $x_0$ is a point of one of these sets, with corresponding value $t_0$ (so $\partial F/\partial t = \partial^2 F/\partial t^2 = 0$ at $(t_0, x_0)$ or $F = \partial F/\partial t = 0$ at $(t_0, x_0)$, respectively), then we work out the type $A_k$ of the singularity which $f = F_{x_0}$ has at $t_0$, by counting the number of derivatives of $f$ which vanish at $t_0$. We then decide whether $F$ versally unfolds $f$ at $t_0$ by finding $\partial F/\partial x_i$ and using the matrix criterion, 6.10p or 6.10. If the criterion is satisfied then locally (near $x_0$) the bifurcation set or discriminant set is diffeomorphic to the standard model applicable to the values of $r$ and $k$ in question. These are listed in chapter 6, (6.16p–18p and 6.16–18).

Lest the reader think that we now have the answer to all questions of this kind, we give several examples where these methods give little or no information. Nevertheless we *do* have the answer to *almost all* questions, as we shall point out from time to time in this chapter, and as we shall make more precise in chapter 9.

### Distance-squared and height functions

7.1 **Distance-squared functions on plane curves**
Let $\gamma: I \rightarrow \mathbb{R}^2$ be a unit speed plane curve, and let $F: I \times \mathbb{R}^2 \rightarrow \mathbb{R}^2$ be
$$F(t, x) = (\gamma(t) - x) \cdot (\gamma(t) - x).$$

Using the results of 2.27 we find that the singular set $S_F$ (6.13p) is $\{(t, \gamma(t) + \lambda N(t) : t \in I, \lambda \in \mathbb{R}\}$ : this is called the *normal bundle* of $\gamma$. Moreover
$$\partial F/\partial t = \partial^2 F/\partial t^2 = 0 \quad \text{iff} \quad \kappa(t) \neq 0 \quad \text{and} \quad x = \gamma(t) + N(t)/\kappa(t)$$
so that the bifurcation set of $F$ is precisely the *evolute* of $\gamma$, that is the locus of centres of curvature.

Next, the condition for $f = F_{x_0}$ to have exactly an $A_2$ at $t_0$ is for $x_0$ to be the centre of curvature at $t_0$ but $\partial^3 F/\partial t^3 (t_0, x_0) \neq 0$. Again referring to 2.27 the condition for this is $\kappa'(t_0) \neq 0$. Likewise the condition for $f$ to have exactly an $A_3$ at $t_0$ is for $x_0$ to be the centre of curvature at $t_0$ and $\kappa'(t_0) = 0$, $\kappa''(t_0) \neq 0$, so that $\gamma$ has an ordinary vertex at $t_0$. Since $r = 2$ here we need not look beyond $A_3$ singularities (higher $A_k$ could never be (p)versally unfolded by $F$).

In order to apply the results of chapter 6 we need to decide whether $F$ is a versal unfolding of $f$ at $t_0$. Now
$$F(t, x) = (X - x_1)^2 + (Y - x_2)^2$$
where $\gamma(t) = (X(t), Y(t))$ – pardon the odd mixture of notation! Thus
$$\partial F/\partial x_1 = -2(X - x_1); \text{ 2-jet at } t_0 = -2(t X'(t_0) + \tfrac{1}{2} t^2 X''(t_0)),$$
$$\partial F/\partial x_2 = -2(Y - x_2); \text{ 2-jet at } t_0 = -2(t Y'(t_0) + \tfrac{1}{2} t^2 Y''(t_0)).$$
The condition for (p)versal unfolding is as follows – see 6.10p.

(i) When $f$ has $A_2$ at $t_0$, we require the $1 \times 2$ matrix $(-2X'(t_0), -2Y'(t_0))$ to have rank 1, which it always does since $\gamma$ is regular.

(ii) When $f$ has $A_3$ at $t_0$, we require the $2 \times 2$ matrix
$$\begin{pmatrix} -2X'(t_0) & -2Y'(t_0) \\ -X''(t_0) & -Y''(t_0) \end{pmatrix}$$
to be nonsingular. But this just says $\kappa(t_0) \neq 0$, which is true since $\partial F/\partial t = \partial^2 F/\partial t^2 = 0$ at $(t_0, x_0)$.

Hence the (p)versal unfolding conditions are automatically satisfied and we deduce:

## 7.2    Proposition

*Let $x$ be a point of the evolute of a (regular) plane curve, $x$ being the centre of curvature at $t$. Then, locally at $x$, the evolute is*

(i) *diffeomorphic to a line in $\mathbb{R}^2$ if the curve does not have a vertex at $t$;*

(ii) *diffeomorphic to an ordinary cusp in $\mathbb{R}^2$ if the curve has an ordinary vertex at $t$.*    □

## 7.3    Remark

At a higher vertex we can deduce nothing from the above method. However we shall see in chapter 9 that, in general, curves do not have any higher vertices (see 9.7). Hence, *for a 'general' plane curve the evolute is locally smooth (like the image of a regular curve) except for the presence of ordinary cusps corresponding with ordinary vertices of the curve.*

Note that the 'cuspidal tangent' in (*ii*), i.e. the limit of tangents at nearby regular points, will be along the normal to $\gamma$ at $t$. For the tangent to the evolute at a regular point is the normal to $\gamma$ (see 2.30(3)), and the result now follows by continuity.

As an example, the evolutes of parabola and ellipse have respectively one and four ordinary cusps: compare figs. 1.5, 2.2. At all other points the evolutes are locally smooth, i.e. parametrized 1-manifolds in $\mathbb{R}^2$ and in particular (images of) regular curves. Compare 2.30(3).

## 7.4 Exercises

(1) Suppose that $\gamma$ has an ordinary vertex at $t_0$ (so $\kappa'(t_0)=0$, $\kappa''(t_0)\neq0$) and that $\kappa(t_0)>0$. Show that the cusp on the evolute points 'towards' $\gamma(t_0)$ when $\kappa$ has a maximum at $t_0$ and 'away' from $\gamma(t_0)$ when $\kappa$ has a minimum. Check this with parabola and ellipse.

(2) The family of potential functions $V$ of chapter 1 is given by the distance from $x \in \mathbb{R}^2$ to the tangent to $\gamma$ at $t$. This distance is (for any $\gamma$) $(x-\gamma(t))\cdot N(t)$. So take $\gamma$ unit speed and define $F(t,x)=(x-\gamma(t))\cdot N(t)$. Assume $\kappa(t)$ is never zero. Show that the bifurcation set is the evolute of $\gamma$ (compare 2.28(2)). Show that $F$ (p)versally unfolds $A_2$ and $A_3$ singularities as in 7.1. What happens if $\kappa$ is sometimes zero?

(3) Let $\gamma$ be unit speed with $\kappa(t)$ never zero, and define $F(t,x)=(x-\gamma(t))\cdot T(t)$. Show that $\partial F/\partial t=0$ is the equation of the line through the centre of curvature of $\gamma$ at $t$ parallel to the tangent at $t$. Thus the bifurcation set of $F$ can also be considered as the envelope of these lines. Show that the family $F$ always (p)versally unfolds the functions $F_x$ when these have $A_2$ or $A_3$ singularities. Show that the condition for $F_x$ to have type $A_2$ at $t$ is

$$x=\gamma + N/\kappa + \kappa'T/\kappa^3$$

where $\kappa\kappa''\neq3\kappa'^2$ (all at $t$, of course). What does this say about the local structure of the bifurcation set?

## 7.5 Distance-squared functions on a space curve

Let $\gamma: I\to\mathbb{R}^3$ be unit speed, with $\kappa(t)$ never zero, and define $F(t,x)=(\gamma(t)-x)\cdot(\gamma(t)-x)$. Then (see 2.34) the singular set $S_F$ is all points of normal planes to $\gamma$, $S_F=\{(t,\gamma(t)+\lambda N(t)+\mu B(t)): t\in I,\lambda,\mu\in\mathbb{R}\}$. This is the *normal bundle* of $\gamma$. The bifurcation set $\mathcal{B}_F$, which can be thought of as the envelope of these planes (see 2.34) consists of points

$$x=\gamma + N/\kappa + \mu B,$$

where $\mu$ is arbitrary. The condition for $F_x$ to have exactly an $A_2$ at $t$ is that $x$ as above is not the centre of spherical curvature at $t$, or $\tau=0$ while

$\kappa \neq 0$ (again see 2.34). The conditions for exactly an $A_3$ or $A_4$ involve complicated expressions in the derivatives of $\kappa$ and $\tau$. It is, however, not hard to find the conditions for $F$ to (p)versally unfold $F_x$ and this is done in the exercises below. Thus (by 6.16p–18p with $r=3$) the local structure of the bifurcation set is smooth, a cuspidal edge or a swallowtail whenever $F_x$ has exactly $A_2$, $A_3$ or $A_4$ (respectively) at $t$ and the conditions in (3) below hold.

### 7.6    Exercises

(1) Writing $\gamma(t)=(X(t), Y(t), Z(t))$ show that $\partial F/\partial x_1 = -2X + 2x_1$ with similar formulas for $\partial F/\partial x_2$ and $\partial F/\partial x_3$.

(2) To find the 3-jets of $\partial F/\partial x_i(t, x_0)$ at $t_0$ (they are independent of $x_0$) take $\gamma$ in standard form with $\gamma(t_0)=0$, $T(t_0)=(1,0,0)$, $N(t_0)=(0,1,0)$, $B(t_0)=(0,0,1)$. Using the result of 2.36(4) show that the 3-jets are then $(-2)$ times

$$t - \tfrac{1}{6}\kappa^2 t^3, \; \tfrac{1}{2}\kappa t^2 + \tfrac{1}{6}\kappa' t^3, \; \tfrac{1}{6}\kappa\tau t^3$$

respectively, where $\kappa$, $\kappa'$, $\tau$ are evaluated at $t_0$.

(3) Deduce that $A_2$ and $A_3$ singularities are always (p)versally unfolded by $F$ and that an $A_4$ singularity at $t_0$ is (p)versally unfolded provided $\tau(t_0)\neq 0$.

(4) **Height functions on a space curve.**    Here we have the family $H: I \times S^2 \to \mathbb{R}$, $H(t, u) = \gamma(t) \cdot u$, where $\gamma$ is a unit speed space curve and $S^2$ is the set of unit vectors in $\mathbb{R}^3$ (so $S^2$ is the unit 2-sphere at the origin of $\mathbb{R}^3$). Here $r=2$, and we need to take local coordinates on part of $S^2$ in order to fit exactly into the general discussion.

(i) (Compare 2.35.) Show that the singular set $S_H$ consists of all points $(t, u)$ with $u$ normal to the curve at $t$. Show also that the bifurcation set of $H$ consists of the points $\pm B(t)$ for $t \in I$: that is $\mathscr{B}_H$ is the set of unit binormal vectors of $\gamma$, together with their reflexions in 0. Thus the projection $\pi: S_H \to S^2$ folds at points over these vectors.

(ii) Show that $H_u$ has type $A_2$ at $t$ iff $u = \pm B(t)$ and $\tau(t) \neq 0$; $A_3$ at $t$ iff $u = \pm B(t)$, $\tau(t) = 0$ and $\tau'(t) \neq 0$.

(iii) To investigate (p)versal unfoldings take $\gamma$ in standard position as in (2) above and use $(x_1, x_2) \mapsto (x_1, x_2, \sqrt{(1 - x_1^2 - x_2^2)})$ as a local parametrization of $S^2$ near $(0, 0, 1)$ so that $H$ can be written

$H: I \times \mathbb{R}^2, (t_0, 0) \to \mathbb{R}$

where

$H(t, x) = \gamma(t) \cdot (x_1, x_2, \sqrt{(1 - x_1^2 - x_2^2)})$.

Find $\partial H/\partial x_1$, $\partial H/\partial x_2$ and show that $H$ always (p)versally unfolds an $A_2$ or an $A_3$ singularity of $H_0$ at $t_0$. What does this show about the bifurcation set of $H$?

As always (see 2.35), height functions have to do with contact with planes. Osculating planes have at least 3-point contact with the curve, but this rises to at least 4-point contact at points of zero torsion. The bifurcation set of $H$ (fig. 7.1) brings this out in a very visible form, by having a cusp at the latter points, provided they are non-degenerate ($\tau' \neq 0$). The singular set $S_H$ is called the *unit normal bundle* of $\gamma$ and the projection $S_H \to S^2$ the *Gauss map* of $\gamma$. There is a similar map for smooth surfaces in $\mathbb{R}^3$ which is of great significance in the study of surfaces. See chapter 11 and Banchoff, Gaffney and McCrory (1982).

We now turn to the 'discriminant' situation, where we consider a family of functions which are not to be regarded as potential functions. Thus we speak of versal unfoldings (not (p)versal), and use the criterion 6.10 to check versality. We are interested in the local structure of discriminant sets $\mathscr{D}_F = \{x: \text{for some } t, \, F = \partial F/\partial t = 0 \text{ at } (t, x)\}$, which for versal unfoldings of $A_1$, $A_2$, $A_3$, are described in 6.16–18.

## 7.7 Height functions on a plane curve; duals

This is back to potential functions, but actually we consider the following slight variant. Let $\gamma: I \to \mathbb{R}^2$ be unit speed and consider

$$\tilde{H}: I \times S^1 \times \mathbb{R} \to \mathbb{R}$$

given by

$$\tilde{H}(t, u, v) = H(t, u) - v = \gamma(t) \cdot u - v.$$

**Fig. 7.1.** Space curve (intersection of a sphere and a cylinder) and the bifurcation set of $H$ (binormal image)

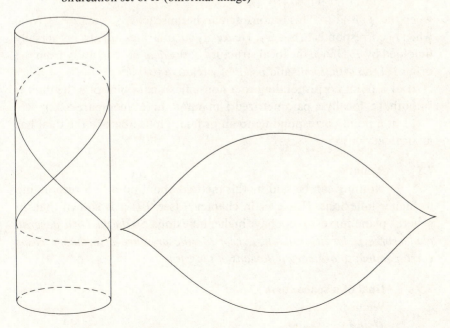

Here $S^1$ is the unit vectors in $\mathbb{R}^2$. We shall need to replace $S^1$ by $\mathbb{R}$, or take local coordinates on $S^1$, in order to be exactly in the general setting. With $S^1$ this is no problem since we can use

$$F: I \times \mathbb{R}^2 \to \mathbb{R}$$

where

$$F(t, x) = \gamma(t) \cdot (\cos x_1, \sin x_1) - x_2.$$

The discriminant set of $\tilde{H}$ is $\{(\pm N(t), \pm \gamma(t) \cdot N(t)): t \in I\}$, which (compare 5.32) is the dual of $\gamma$ together with its reflexion obtained by changing the sign in both $S^1$ and $\mathbb{R}$. Using $F$ we can think of the dual in $\mathbb{R}^2$, or we can project the discriminant of $\tilde{H}$ to $\mathbb{R}^2$ as in 5.32, thereby confusing things about the origin.

Since $r = 2$ we need only consider $A_1$ and $A_2$ singularities. Now $\tilde{H}_u$ has $A_1$ at $t$ iff $u = \pm N(t)$ and $\kappa(t) \neq 0$; $A_2$ at $t$ iff $u = \pm N(t)$ and $\kappa(t) = 0$, $\kappa'(t) \neq 0$ (compare 2.29).

For investigating versality we use $F$. Thus

$$\partial F/\partial x_1 = \gamma(t) \cdot (-\sin x_1, \cos x_1); \ \partial F/\partial x_2 = 1.$$

In view of the 1, $A_1$ singularities are certainly versally unfolded, for the matrix of 6.10 has the form (? 1) which has rank 1.

As for $A_2$, we need to consider the matrix whose columns are the 1-jets with constant of $\gamma(t) \cdot (-\sin x_1, \cos x_1)$ and 1. This matrix is

$$\begin{pmatrix} \gamma(t_0) \cdot (-\sin x_1, \cos x_1) & 1 \\ \gamma'(t_0) \cdot (-\sin x_1, \cos x_1) & 0 \end{pmatrix}$$

when the $A_2$ is at $t_0$. This is nonsingular, because $(\cos x_1, \sin x_1) = \pm N(t_0)$ when $t_0$ corresponds to $x \in \mathscr{D}_F$. Hence $A_2$ singularities are always versally unfolded by $F$. Hence the local structure of the *dual* of $\gamma$, or, away from the origin, of the *orthotomic* and *pedal* of $\gamma$ relative to 0 is:

(*i*) at a point corresponding to a non-inflexional point of $\gamma$ the dual is smooth, i.e. locally a parametrized 1-manifold in $\mathbb{R}^2$ (compare 5.32);

(*ii*) at a point corresponding to an ordinary inflexion of $\gamma$ the dual has an ordinary cusp.

**7.8    Remark**

Nothing can be said by this method about points corresponding to higher inflexions. However in chapter 9 (see 9.7) it is shown that, in general, plane curves do not have higher inflexions. So *the dual of a 'general' plane curve is locally smooth, except for the presence of ordinary cusps corresponding to ordinary inflexions of the curve.*

**7.9    Dual of a space curve**

We use

$$\tilde{H}: I \times S^2 \times \mathbb{R} \to \mathbb{R}$$

given by
$$\tilde{H}(t, u, v) = H(t, u) - v = \gamma(t) \cdot u - v,$$
where $\gamma$ is a unit speed space curve and $S^2$ is the set of unit vectors in $\mathbb{R}^3$. Now an oriented plane in $\mathbb{R}^3$ is specified by giving a unit vector $u$ and a number $v$; the equation of the plane is then $x \cdot u = v$ and $(u, v)$ and $(-u, -v)$ give the same plane with opposite orientations. In this way the oriented planes in $\mathbb{R}^3$ correspond one-to-one with the points of $S^2 \times \mathbb{R}$.

Given $t \in I$ we can specify an *oriented tangent plane* to $\gamma$ at $t$, i.e. an oriented plane containing the tangent line at $t$, by a unit vector $u$ perpendicular to $T(t)$: the plane is then the one with normal vector $u$ and passing through $\gamma(t)$ (fig. 7.2) so it has equation $x \cdot u = \gamma(t) \cdot u$. The set of oriented tangent planes to $\gamma$, called the *dual* of $\gamma$, is then identified with $\{(u, v) \in S^2 \times \mathbb{R} : v = \gamma(t) \cdot u, \ T(t) \cdot u = 0\}$, which is precisely the discriminant set of $\tilde{H}$.

Let $(u, v) \in \mathcal{D}_{\tilde{H}}$, with corresponding value $t$, so that $u = \lambda N(t) + \mu B(t)$ for some $\lambda, \mu \in \mathbb{R}$ and $v = \gamma(t) \cdot u$. The function $\tilde{H}_{(u,v)}$ has

$A_1$ at $t$    iff    $\lambda \neq 0$

$A_2$ at $t$    iff    $\lambda = 0, \ \tau(t) \neq 0$ (so $\mu = \pm 1$)

$A_3$ at $t$    iff    $\lambda = 0, \ \tau(t) = 0, \ \tau'(t) \neq 0$.

### 7.10    Exercise

By putting $\gamma$ in standard position as in 7.6(2) and using
$$F: I \times \mathbb{R}^3 \rightarrowtail \mathbb{R}$$
given by
$$F(t, x) = \gamma(t) \cdot (x_1, x_2, \sqrt{(1 - x_1^2 - x_2^2)}) - x_3$$
(compare 7.6(4)) show that $F$ always versally unfolds $A_k$ singularities of $F_0$ for $k \leq 3$. Deduce that the local structure of the dual of $\gamma$ is

**Fig. 7.2.** A tangent plane to a space curve

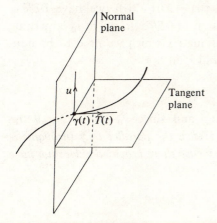

(*i*) smooth (a parametrized 2-manifold in $\mathbb{R}^3$) at points of the dual corresponding to tangent planes other than the osculating plane;

(*ii*) cuspidal edge at points of the dual corresponding to osculating planes at points of non-vanishing torsion;

(*iii*) swallowtail at such points where $\tau = 0$ but $\tau' \neq 0$. (What is the geometrical interpretation of the curve of self-intersection on the swallowtail?) Again these are the only possibilities for a 'general' space curve – see 9.10.

## Envelopes

We are still in the 'discriminant' situation, and recognize versal unfoldings by 6.10.

### 7.11    Evolute of a plane curve as envelope of normals

This has already been considered in 5.28 except at points of regression. For a unit speed $\gamma: I \to \mathbb{R}^2$ we use $F: I \times \mathbb{R}^2 \to \mathbb{R}$ given by $F(t, x) = (x - \gamma(t)) \cdot T(t)$ so that the discriminant $\mathcal{D}_F$ is the evolute of $\gamma$. The function $f$ given by $f(t) = F(t, x_0)$ has type $A_2$ at $t_0$ iff $x_0$ is the centre of curvature of $\gamma$ at $t_0$ and $\kappa'(t_0) = 0$, $\kappa''(t_0) \neq 0$, i.e. $\gamma$ has an ordinary vertex at $t_0$. The calculation to show that $F$ versally unfolds $f$ is almost indistinguishable from that in 7.1, so we arrive at the same result: *the evolute has an ordinary cusp corresponding to an ordinary vertex of* $\gamma$. (Compare also 3.13(5).)

### 7.12    Parallels

Here $\gamma: I \to \mathbb{R}^2$ is a unit speed plane curve and $F(t, x) = (x - \gamma(t)) \cdot (x - \gamma(t)) - r^2$ where $r > 0$. The points of regression were found in 5.30, and are points $x_0 = \gamma(t_0) + N(t_0)/\kappa(t_0)$ where $r = \pm 1/\kappa(t_0)$, whichever is $> 0$. The condition for $f$ $(f(t) = F(t, x_0))$ to have exactly $A_2$ at $t_0$ (i.e. $\partial^3 F/\partial t^3(t_0, x_0) \neq 0$) is $\kappa'(t_0) \neq 0$, since $F$ has the same derivatives as the distance-squared function. Writing $\gamma(t) = (X(t), Y(t))$, we have $\partial F/\partial x_1 = 2(x_1 - X(t))$ so the 1-jet with constant of $\partial F/\partial x_1(t, x_0)$ at $t_0$ is (writing $x_0 = (a, b)$) $2(a - X(t_0)) - 2tX'(t_0)$. The matrix which we need to be nonsingular to check that $F$ versally unfolds $f$ at $t_0$ is

$$\begin{pmatrix} 2(a - X(t_0)) & 2(b - Y(t_0)) \\ -2X'(t_0) & -2Y'(t_0) \end{pmatrix}.$$

The determinant is $4(x_0 - \gamma(t_0)) \cdot N(t_0)$ and since $x_0 - \gamma(t_0) = N(t_0)/\kappa(t_0)$ this is nonzero. (Compare 3.13(4).) Hence *the parallel to* $\gamma$ *through the centre of curvature at* $t$ *has an ordinary cusp there provided* $\gamma$ *does not have a vertex at* $t$. See fig. 2.21.

It is also of interest to consider the family

$$G: I \times \mathbb{R}^2 \times \mathbb{R}_{>0} \to \mathbb{R}$$

given by

$$G(t, x, r) = (x - \gamma(t)) \cdot (x - \gamma(t)) - r^2.$$

The discriminant $\mathscr{D}_G$ is

$$\{(x, r): \text{ there exists } t \text{ with } x = \gamma(t) \pm r N(t)\},$$

that is $\mathscr{D}_G$ consists of all the parallels stacked up to form a surface in $\mathbb{R}^3$ – or two surfaces, given by $+r$ and $-r$. The condition for $G_{(x_0, r_0)}$ to have type $A_k$ at $t_0$ is the same as the condition for $F_{x_0}$ to have type $A_k$ at $t_0$, and it is easy to verify that $G$ always satisfies the versal unfolding conditions for types $A_{\leqslant 3}$. Hence the surface of parallels is

(i) smooth near a point $(x_0, r_0)$ corresponding to $t_0$ with $x_0$ not the centre of curvature of $\gamma$ at $t_0$,

(ii) cuspidal edge when $x_0$ is the centre of curvature at $t_0$, $r_0 = |1/\kappa(t_0)|$ and $\gamma$ does not have a vertex at $t_0$,

(iii) swallowtail when $x_0$ is the centre of curvature of $\gamma$ at an ordinary vertex $t_0$ and $r_0 = |1/\kappa(t_0)|$.

The *sections* of $\mathscr{D}_G$ by planes $r = $ constant are the individual parallels (or pairs of parallels). Notice that the sections of the swallowtail described in 5.36 correspond closely with the evolving parallels of a parabola near to the centre of curvature at the vertex, as in fig. 2.21. (In fact in case (iii) the parallels will always evolve in this way, but we cannot prove this here. (The methods for proving this were developed by Arnold (1976)).) Notice also that, in case (iii), projecting $\mathscr{D}_G$ to the $\mathbb{R}^2$ factor the image of the cuspidal edge on the swallowtail is precisely the evolute of $\gamma$ near the point $x_0$, the swallowtail point $(x_0, r_0)$ corresponding to the cusp $x_0$ on the evolute.

## 7.13    Apparent contours of surfaces

In 5.4(3) we took $F(t, x_1, x_2)$ with 0 a regular value of $F$, and considered the 'apparent contour' of $M = F^{-1}(0)$ in the $t$-direction, which is the envelope of the 'smooth part' of the curves $F_t = 0$, that is the envelope of $F$ restricted to $\{(t, x): \partial F/\partial x_1 \text{ or } \partial F/\partial x_2 \neq 0\}$.

A point of $M$ where $F = \partial F/\partial t = 0$ but $\partial^2 F/\partial t^2 \neq 0$ is called a *fold point* (or more fully an ordinary fold point) of the projection $\pi: M \to \mathbb{R}^2$, $\pi(t, x_1, x_2) = (x_1, x_2)$. The projection of such a fold point is a regular point of the envelope (5.26). A point of $M$ where $F = \partial F/\partial t = \partial^2 F/\partial t^2 = 0$ but $\partial^3 F/\partial t^3 \neq 0$ is called a *cusp point* of $\pi$, and its projection is a point of regression on the envelope. The function $F_{x_0}$ clearly has an $A_1$ singularity at $t_0$ when $(t_0, x_0)$ is a fold point, and an $A_2$ when $(t_0, x_0)$ is a cusp point.

The condition for $F$ to versally unfold $F_{x_0}$ (6.10) is always satisfied at a fold point and works out to

$$\frac{\partial F}{\partial x_1} \frac{\partial^2 F}{\partial t \partial x_2} \neq \frac{\partial F}{\partial x_2} \frac{\partial^2 F}{\partial t \partial x_1}$$

at a cusp point.

The results of chapter 6 now assert that provided these conditions are satisfied the projection 'looks like' the projections of fig. 6.3 for the fold and fig. 1.6 for the cusp; see also fig. 7.3. More precisely let $G: \mathbb{R} \times \mathbb{R}^2$, $(0, 0) \to \mathbb{R}^2, 0$ be defined by $G(t, u) = t^2 + u_2$ for the fold case and $t^3 + tu_1 + u_2$ for the cusp case. Then there exists a local diffeomorphism

$$\phi = (a, b): \mathbb{R} \times \mathbb{R}^2, (t_0, x_0) \to \mathbb{R} \times \mathbb{R}^2, (0, 0), \phi(t, x) = (a(t, x), b(x))$$

making the following diagram commute:

$$
\begin{array}{ccc}
M = F^{-1}(0) & \overset{\phi}{\longrightarrow} & G^{-1}(0) \\
\downarrow \pi & & \pi \downarrow \\
\mathbb{R}^2 & \overset{b}{\longrightarrow} & \mathbb{R}^2
\end{array}
$$

where $\pi$ is in each case projection on the last two coordinates. Thus $\pi$: $F^{-1}(0) \to \mathbb{R}^2$ is taken diffeomorphically to $\pi: G^{-1}(0) \to \mathbb{R}^2$, and the latter provides a 'normal form' for projections of surfaces in a neighbourhood of a fold or cusp point. In particular the local structure of the envelope $\mathscr{D}_F$ is respectively smooth (a regular curve) and an ordinary cusp, as in 6.16, 6.17, but we can say more. Parametrizing the surface $G^{-1}(0)$ by $(t, u_1, -t^2)$ for the fold and $(t, u_1, -t^3 - u_1 t)$ for the cusp the projections assume the normal forms, as maps $\mathbb{R}^2, 0 \to \mathbb{R}^2, 0$:

$$(u_1, t) \mapsto (u_1, -t^2) \text{ (fold)}; (u_1, t) \mapsto (u_1, -t^3 - u_1 t) \text{ (cusp)}.$$

It is a theorem of Whitney (see for example Bröcker and Lander (1975), chapter 8) that these are generically the only such smooth maps other than

**Fig. 7.3.** Generic local maps $\mathbb{R}^2 \to \mathbb{R}^2$

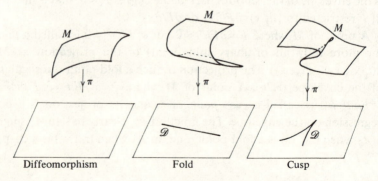

Diffeomorphism          Fold          Cusp

a local diffeomorphism. (Smooth maps $\mathbb{R}^2 \to \mathbb{R}^2$ can be visualized by folding fabrics on a table: compare a pleat or dart with the cusp.)

Here are a couple of examples. Let $F$ be a polynomial of degree 2 in $t$, $F(t, x) = A(x)t^2 + B(x)t + C(x)$, with 0 a regular value of $F$. Then the apparent contour of $F^{-1}(0)$ in the $t$-direction is the set of $x$ for which $B^2 - 4AC = 0$. A point of this locus will be a regular point of the envelope iff $A(x) \neq 0$. Clearly if $A$ does vanish at a point of the apparent contour then $B$ and $C$ will also vanish there. In this unlikely event (3 equations in 2 unknowns having a solution) a whole line in the $t$-direction lies on $M$ and projects to a point. Can you see why a cusp point cannot occur for this $F$?

Secondly consider $F(t, x) = t^3 + f(x)t + g(x)$ (compare 5.5(5)). The apparent contour has an ordinary cusp at $x_0 \in \mathbb{R}^2$ when $f(x_0) = g(x_0) = 0$ and $\partial f/\partial x_1 \cdot \partial g/\partial x_2 \neq \partial f/\partial x_2 \cdot \partial g/\partial x_1$ at $x_0$. This is also the condition for $x \mapsto (f(x), g(x))$ to be a local diffeomorphism at $x_0$, so says more or less that $f$ and $g$ are 'independent' at $x_0$. It can also be interpreted as saying that $f^{-1}(0)$, $g^{-1}(0)$ are, near $x_0$, regular curves having distinct tangents at $x_0$ (crossing transversally, in the language of chapter 8).

Before proceeding we remark that provided there are no points $(t, x)$ satisfying any of the three conditions:

(i) $F = \partial F/\partial t = \partial F/\partial x_1 = \partial F/\partial x_2 = 0$;

(ii) $F = \partial F/\partial t = \partial^2 F/\partial t^2 = \partial^3 F/\partial t^3 = 0$;

(iii) $F = \partial F/\partial t = \partial^2 F/\partial t^2 = \partial F/\partial x_1 \cdot \partial^2 F/\partial t \partial x_2 - \partial F/\partial x_2 \cdot \partial^2 F/\partial t \partial x_1 = 0$,

every point of $F^{-1}(0)$ is a fold point or a cusp point of the projection $\pi$, or $\pi$ is a local diffeomorphism there. But each of the conditions (i)–(iii) consists of four equations in three unknowns, and Sard's theorem would lead us to expect that such systems of equations have no solutions (compare the remarks following 4.18). Thus in general a projection from a surface to $\mathbb{R}^2$ should look locally like one of the pictures in fig. 7.3. See 8.21 where this vague reasoning is made precise for smooth algebraic surfaces.

The result of Whitney referred to above was one of the first in the study of generic smooth maps. The methods we used to obtain normal forms also work in higher dimensions. Suppose that 0 is a regular value of $F: \mathbb{R}^{n+1} \to \mathbb{R}$ so that $M = F^{-1}(0)$ is (near every point) a parametrized $n$-manifold in $\mathbb{R}^{n+1}$. Using coordinates $(t, x_1, \ldots, x_n)$ in $\mathbb{R}^{n+1}$, those parts of the sets $F_t^{-1}(0)$ which consist of regular points of $F_t$ will, as in the case $n = 2$, have an envelope which is the apparent contour of $M$ in the $t$-direction. The family of functions $F_x(t) = F(t, x)$ can be *expected* to have only $A_k$ singularities, $k \leqslant n$, at points where $F(t, x) = 0$, and in general these will be versally unfolded by $F$ (see 6.21). The same argument as above produces normal forms, as maps $\mathbb{R}^n, 0 \to \mathbb{R}^n, 0$ as follows.

$$(u, t) \mapsto (u_1, \ldots, u_{n-1}, -(t^{k+1} + u_1 t^{k-1} + \cdots + u_{k-1}t)), \quad 1 \leqslant k \leqslant n,$$

apart from a local diffeomorphism, which has normal form $(u, t) \mapsto (u, t)$.

These $n$ normal forms are examples of Morin singularities, named after their discoverer Bernard Morin – a mathematician with astonishing geometrical insight, despite his total blindness. It is possible to prove that locally almost all projections of parametrized $n$-manifolds in $\mathbb{R}^{n+1}$ to $\mathbb{R}^n$ have one of these forms – in particular such projections are locally at most $n+1$ sheeted: for each $x \in \mathbb{R}^n$ there are at most $n+1$ values of $t$ with $(t, x) \in F^{-1}(0)$.

The study of the structure of 'almost all' smooth maps $\mathbb{R}^n \to \mathbb{R}^p$ is a fascinating and major part of singularity theory. Most of the known results are due to John Mather, following earlier ideas of Whitney and Thom. See for example Golubitsky and Guillemin (1973), Gibson (1979), Martinet (1982).

## 7.14    Exercises

(1) **Orthotomics** Take $F(t, x) = x \cdot x - 2\gamma(t) \cdot x$ as in 5.32, where $\gamma$ is a unit speed curve in the plane, with $\gamma(t)$ never 0. The envelope (discriminant) of $F$ is then $\{0\}$ together with the orthotomic of $\gamma$ relative to 0. The points of regression are 0 and $x = 2(\gamma(t) \cdot N(t))N(t)$ where $\kappa(t) = 0$. Show that in the latter case $F_x$ has exactly $A_2$ at $t$ provided $\gamma(t) \cdot N(t) \neq 0$ and $\kappa'(t) \neq 0$. (Since $F_0$ is identically zero it does not have an $A_k$, for any $k$, at any $t$.) Show that $F$ versally unfolds any such $A_2$ singularity (the calculation is very similar to 7.12 above). Deduce that the orthotomic has an ordinary cusp at $x = 2(\gamma(t) \cdot N(t))N(t)$ whenever the tangent to $\gamma$ at $t$ does not pass through the origin and $\gamma$ has an ordinary inflexion at $t$. Compare the results on duals in 7.7, recalling that the pedal or orthotomic give an accurate picture of the dual (up to diffeomorphism) away from 0 (see 5.32).

(2) **Antiorthotomics** Let $\gamma: I \to \mathbb{R}^2$ be unit speed where $\gamma(t)$ is never 0 and $\gamma(t) \cdot N(t)$ is never 0 (i.e. no tangent to $\gamma$ passes through the origin). We seek a curve $\beta$ whose orthotomic is $\gamma$ and bearing in mind the orthotomic construction as the locus of reflexions of 0 in the tangent lines to $\gamma$ (5.32) we consider the envelope of perpendicular bisectors of lines joining $\gamma(t)$ to 0, as in fig. 7.4 (why is this reasonable?). Show that the appropriate $F$ is $F(t, x) = \gamma(t) \cdot (2x - \gamma(t))$. Show that the envelope consists of points $x = \gamma(t) + \lambda N(t)$ where $\lambda = -\gamma(t) \cdot \gamma(t)/2\gamma(t) \cdot N(t)$. Show that $x$ is a point of regression, with $t$ corresponding to $x$, iff the circle of curvature of $\gamma$ at $t$ passes through 0. Show further that the envelope (i.e. the antiorthotomic of $\gamma$) has an ordinary cusp at such an $x$ provided $\gamma$ does not have a vertex at $t$ (fig. 7.4).

(3) The construction in (2) can be done by **paper-folding** (fig. 7.4). Draw $\gamma(I)$ on paper and, selecting $t$, fold the paper so that $\gamma(t)$ coincides with a fixed point 0 on the paper. The resulting crease is the perpendicular bisector (why?), so doing this for lots of values of $t$ gives a good picture of the anti-orthotomic as an envelope of lines. Try this with a circle for $\gamma(I)$ and 0 (*i*) inside, (*ii*) outside the circle. What feature of the antiorthotomic corresponds to the tangents through 0 in (*ii*)? What happens if 0 is on the circle?

(4) **Envelope of tangents to a plane curve** As in 5.29(1) let $\gamma: I \to \mathbb{R}^2$ be unit speed and $F(t, x) = (x - \gamma(t)) \cdot N(t)$. The points of regression of the envelope

**Fig. 7.4.** An antiorthotomic of an ellipse

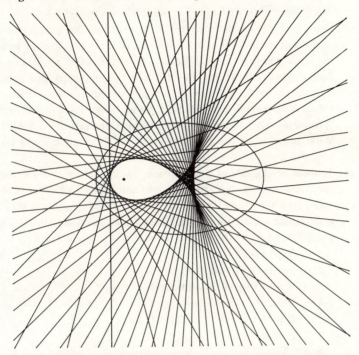

of $F$ are (*i*) points $x_0 = \gamma(t_0)$ where $\kappa(t_0) = 0$, $\kappa'(t_0) \neq 0$; (*ii*) points $x_0 = \gamma(t_0) + \lambda T(t_0)$ ($\lambda$ an arbitrary real number) when $\kappa(t_0) = \kappa'(t_0) = 0$. Show that in (*i*) $f = F_{x_0}$ has type $A_2$ at $t_0$ while in (*ii*) it has type $A_{\geqslant 3}$. Show that $F$ is *never* a versal unfolding of $f$ in (*i*). What does this *not* tell you about the local structure of the envelope? Are you surprised? (If you are, look again at 5.5(2).)

(5) **Envelope of polar lines** Let $\gamma: I \to \mathbb{R}^2$ be unit speed. The *polar line* of $\gamma(t)$ with respect to the unit circle (centre (0, 0)) is the line $x \cdot \gamma(t) = 1$ in the plane. (For $\gamma(t)$ outside the circle it joins the points of contact of tangents to the circle from $\gamma(t)$ – can you see why?) Investigate the envelope of polar lines. Do the methods of chapter 6 give any information here?

(6) For the **embroidery** example of 5.7(2) the only point of regression is $x_0 = (-\frac{1}{3}, 0)$, given by $t_0 = \pi$. (Points $(1, x_2)$, $t = 0$, do not count since $F_0$ does not have 0 as a regular value.) Show that $F$ is a versal unfolding of $f$ ($f(t) = F(t, x_0)$) at $t = \pi$. Hence the envelope has an ordinary cusp at $(-\frac{1}{3}, 0)$. (For the envelope where $2t$ is replaced by $mt$ we find $m - 1$ ordinary cusps, corresponding to $(m-1)t = (2n+1)\pi$, $n$ an integer. The envelope is parametrized $(m+1)x_1 = m\cos t + \cos mt$, $(m+1)x_2 = m\sin t + \sin mt$, where $(m-1)t = 2n\pi$ has to be excluded because 0 is not a regular value of these $F_t$.)

(7) More embroidery. Let

$F(t, x) = x_1 \sin t + x_2(1 + \cos t) - (\sin t + \sin 2t)$, $-\pi < t < \pi$.

Show that $F = 0$ is the equation of the chord joining points $(\cos \theta, \sin \theta)$ of the unit circle for $\theta = t$ and $\theta = \pi - 2t$. (For $t = \pm \frac{1}{3}\pi$ it is the equation of the tangent at $(\cos t, \sin t)$.) We have: $F = \partial F/\partial t = 0$ iff $x = (2 \cos t + \cos 2t, 2 \sin t - \sin 2t)$; only verify this if you find trigonometry irresistible. Show that the envelope of $F$ has three points of regression $x_0$, corresponding values of $t_0$ being $0$, $\pm \frac{2}{3}\pi$. Find the points $x_0$ and show that in each case $f$ $(f(t) = F(t, x_0))$ has type $A_2$ at $t_0$. Which are versally unfolded by $F$? What does this tell you about the envelope? Draw a picture by taking say $t = k\pi/18, k = -17, -16, \ldots, 17$.

(8) **Tangent developable of a space curve** As in 5.34 let $\gamma: I \to \mathbb{R}^3$ be a unit speed space curve with $\kappa(t)$ never zero and define $F: I \times \mathbb{R}^3 \to \mathbb{R}$ by $F(t, x) = (x - \gamma(t)) \cdot B(t)$. The points of regression have the form $x = \gamma(t) + \lambda T(t) + \mu N(t)$ where $\lambda = \mu = 0$ when $\tau(t) \neq 0$, and $\mu = 0$ when $\tau(t) = 0$, $\tau'(t) \neq 0$. Show the following about the singularity of $F_x$ at $t$ for such $x$ and $t$:

   (i) $A_2$ if $\tau(t) \neq 0$,
   (ii) $A_2$ if $\tau(t) = 0$, $\tau'(t) \neq 0$, $\lambda \neq 0$,
   (iii) $A_2$ if $\tau(t) = \tau'(t) = 0$, $\tau''(t) \neq 0$, $\mu \neq 0$,
   (iv) $A_3$ if $\tau(t) = 0$, $\tau'(t) \neq 0$, $\lambda = 0$,
   (v) $A_3$ if $\tau(t) = \tau'(t) = 0$, $\tau''(t) \neq 0$, $\mu = 0$, $\lambda \neq 0$,
   (vi) $A_{\geqslant 4}$ otherwise.

   Show that only in case (i) does $F$ give a versal unfolding of $F_x$ at $t$. (Hint: the $\partial F/\partial x_i$ are just the components of $B$, and $B' = 0$ when $\tau = 0$.) What does this say about the structure of the tangent developable at points on the image of $\gamma$ where the torsion does not vanish? (Compare fig. 7.5, which shows half of the tangent developable.)

(9) Investigate the **envelope of normal planes** of a space curve, i.e. consider $F(t, x) = (x - \gamma(t)) \cdot T(t)$, where $\gamma$ is a unit speed space curve. The calculations are virtually identical with those of 7.5 and 7.6.

## Some different local pictures

It is also possible to find local forms of the 'envelope' $E_3$ of a family $F(t, x) = 0$ of curves, as discussed in 5.16: it is defined as the boundary of the union of the curves $F_t(x) = 0$. Since $E_3$ is defined in terms of those $x$ for which $F(t, x) = 0$ has a solution for $t$, it is easy to check that $F(t, x)$ and $F(a(t, x), b(x))$, with $b$ a local diffeomorphism, will give locally diffeomorphic sets $E_3$. Thus by an argument similar to 6.14p any two $r$-parameter versal unfoldings of functions of the same type $A_k$ will have locally diffeomorphic sets $E_3$. Examining the standard examples 6.16 and 6.17 with $r = 2$ we find:

   (i) For the fold $(A_1)$, $E_3$ is locally a smooth curve,
   (ii) For the cusp $(A_2)$, $E_3$ is empty.

**Fig. 7.5.** Tangent developable of a helix

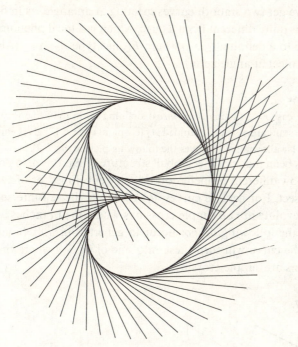

**Fig. 7.6.** Global structure of $E_3$

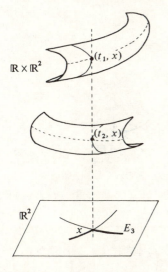

Thus, with $r = 2$ (family of curves), provided $F$ is always a versal unfold-ing of all the singularities of $F_x$ for all $x$, the set $E_3$ has the *local* structure of a smooth curve (or is empty). Note, however, that two folds of the surface $F(t, x) = 0$ in $\mathbb{R}^3$ may project to overlapping regions of $\mathbb{R}^2$, so that when we take into account the 'global' structure of $E_3$ there may be points where we get two smooth curves meeting at an angle, as in fig. 7.6. Note that this is quite different from a cusp, which is a local phenomenon, with $(t, x)$ close to a definite $(t_0, x_0)$ where $F_{x_0}$ has type $A_2$ at $t_0$. (Also the arms of a cusp meet at angle zero.)

**7.15    Exercise**

For the case $r = 3$ (family of *surfaces*), defining $E_3$ in the same way, show that the standard versally unfolded $A_2$ (cuspidal edge, 6.17) gives empty $E_3$ and $A_3$ (swallowtail, 6.18) gives the following picture (fig. 7.7).

Mention has been made of the 'global' structure of $E_3$. It is also possible to determine how different branches of ordinary envelopes, or of evolutes or duals, intersect. For generic curves, for example, the self-intersections of the dual or evolute are all simple crossings where the branches are not tangent. Thus the following do *not* occur in general (fig. 7.8). Small perturbations of the curve will turn these into one of fig. 7.9 where the only self-intersections are simple crossings.

Fig. 7.7.

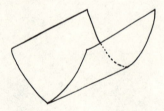

Fig. 7.8.

Fig. 7.9.

## Symmetry sets

Suppose a closed curve $C$ in $\mathbb{R}^2$ has an axis of symmetry, as in fig. 7.10. Then a circle centred on the axis of symmetry and touching $C$ at a point $P$ will also touch $C$ at the symmetric point $P'$. Not every point of the axis is a suitable centre for such a circle; if $C$ has no inflexions (for example, an ellipse) then it is fairly clear that the interval of suitable centres connects the centres of curvature at the points (necessarily vertices of $C$) where the axis meets $C$. (What is the interval for one of the limaçons of fig. 2.12?) It also seems plausible that the envelope of these circles will be $C$ itself – though the behaviour at an endpoint of the interval of centres is unclear. See 7.16(3) below.

There is a local version of this, as follows. Let $F: I \times \mathbb{R}^2 \to \mathbb{R}$ be the family of distance-squared functions on a curve $\gamma: I \to \mathbb{R}^2$. Notice that in the above situation the function $F_x: I \to \mathbb{R}$, where $x$ is an interior point of the interval of centres, has two singularities, type $A_1$ or worse, at say $t_1$ and $t_2$ – the circle touches $C = \gamma(I)$ at $\gamma(t_1)$ and $\gamma(t_2)$ – where $F_x(t_1) = F_x(t_2)$. We say that the singularities of $F_x$ are at the *same level*. For any curve $\gamma$, then, we can consider the *symmetry set*, namely

$$\{x \in \mathbb{R}^2 : \text{there exist } t_1, t_2 \in I \text{ with } \gamma(t_1) \neq \gamma(t_2),$$
$$\partial F/\partial t(t_1, x) = \partial F/\partial t(t_2, x) = 0, \ F(t_1, x) = F(t_2, x)\}.$$

This simply means that there is a circle, centre $x$, touching $\gamma(I)$ at two distinct points. The tangents at these two points are symmetric about a suitable line through $x$ (fig. 7.11).

**Fig. 7.10.** Axis of symmetry

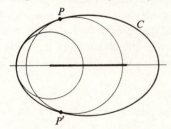

**Fig. 7.11.** Infinitesimal axis of symmetry

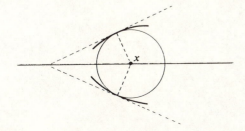

The symmetry set of an ellipse consists of two segments (fig. 7.12, left). The end-points are not part of the symmetry set: the two points of contact of the circle coincide there, and $F_x$ has a single $A_3$ singularity at the appropriate value of $t$. As $x$ approaches an end-point the two $A_1$ singularities of $F_x$ coalesce into this $A_3$, and the circle has 4-point contact with the ellipse. The right-hand diagram of fig. 7.11 shows the $A_2$-set of $F$ (evolute of the ellipse) near an $A_3$ point and the symmetry set ($A_1 A_1$ set) close by.

### 7.16    Exercises

(1) Let $F(t, x) = \frac{1}{4}t^4 + \frac{1}{2}x_1 t^2 + x_2 t$. Show that the $A_1 A_1$ set (those $x$ for which $F_x$ has two $A_1$ singularities at the same level) is precisely $\{x: x_2 = 0, \ x_1 < 0\}$. The $A_2$ set here is the usual cusp $4x_1^3 + 27x_2^2 = 0$ with $A_3$ at $x_1 = x_2 = 0$. It will be the same whenever the family of distance-squared functions is a (p)versal unfolding of an $A_3$ singularity. So the symmetry set always has a branch ending at an ordinary cusp ($A_3$ point) on the evolute ($A_2$ set).

(2) Consider the ellipse $\gamma(t) = (a \cos t, b \sin t)$ with $a > b > 0$. Let $F(t, x)$ $= (x_1^2 + x_2^2)a^2 - 2x_1 a(a^2 - b^2)\cos t + a^2(a^2 - b^2)\cos^2 t - a^2 b^2$. Show that $F(t, x) = 0$ is the equation of a circle centred on the $x_1$-axis and touching the ellipse at the points $\gamma(t)$ and $\gamma(-t)$ (except for $t = 0$ and $t = \pi$ when the circle has 4-point contact with the ellipse at $\gamma(t)$). Show that the envelope of this family of circles consists of the ellipse together with the circles of curvature at $(a, 0)$ and $(-a, 0)$. (Harder: prove that the 'envelope' $E_1$ as defined by intersections of nearby circles (see 5.8) is precisely the ellipse.)

(3) Let $f$ be a smooth *even* function of the real variable $t$. Assume the theorem that there exists a smooth function $g$ with $f(t) = g(t^2)$. (This apparently innocent result is actually hard to prove: see for example Martinet (1982).) Consider the graph $x_2 = f(x_1)$, parametrized $\gamma(t) = (t, g(t^2))$. Assume $f''(0) \neq 0$, which clearly implies $g'(0) \neq 0$. Find the point $(0, \theta(t))$ where the normal at $\gamma(t)$ meets the $x_2$-axis and hence find the equation of the circle centre $(0, \theta(t))$ touching the graph at $(t, g(t^2))$ and $(-t, g(t^2))$ (except for at least

**Fig. 7.12.** Symmetry set of an ellipse, and symmetry set near an $A_3$ point

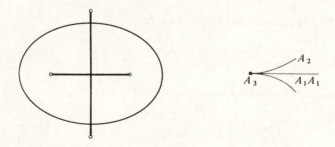

4-point contact when $t=0$). Show that the envelope of these circles contains the whole of the circle of curvature at $(0, g(0))$.

(4) Show that for the family $F(t, x) = t^5 + x_1 t^2 + x_2 t$ (where $F_0$ has type $A_4$ at $t=0$) there do not exist any points $x$ for which $F_x$ has two singularities $A_{\geqslant 1}$ at the same level.

To study symmetry sets further we describe them as follows. Let $\gamma$ and $F$ be as before and let

$$\tilde{F}: I \times \mathbb{R}^2 \times \mathbb{R} \to \mathbb{R}$$

be

$$\tilde{F}(t, x, u) = F(t, x) - u.$$

Then the discriminant $\mathscr{D}_{\tilde{F}}$ is

$$\mathscr{D}_{\tilde{F}} = \{(x, u): \text{there exists } t \text{ with } \partial F/\partial t(t, x) = 0, \ F(t, x) = u\}.$$

We know the local structure of this set provided certain versal unfolding conditions are satisfied (compare 6.14), and furthermore for 'general' curves they will be. See exercise (1) below. Now $\mathscr{D}_{\tilde{F}}$ is the projection to $\mathbb{R}^2 \times \mathbb{R}$ of the set of $(t, x, u)$ with $\partial F/\partial t(t, x) = 0$ and $F(t, x) = u$. Consider the set of $(x, u)$ for which there are *two* points $(t_1, x, u)$, $(t_2, x, u)$ over $(x, u)$, the set formed by the self-intersections of $\mathscr{D}_{\tilde{F}}$. The diagram (fig. 7.13) suggests this with one dimension removed. In our case the self-intersection set will consist of *curves*. There are four cases. (*i*) The swallowtail, which arises as the discriminant of a versal unfolding $\tilde{F}$ of an $A_3$ singularity (see 6.18) already has a self-intersection. The other cases come from planes and cuspidal edges intersecting each other, as in fig. 7.14.

Finally, the symmetry set is obtained from the self-intersection set by a further projection $(x, u) \mapsto x$. We obtain the following pictures (fig. 7.15), where the labelling marks the singularities of the distance-squared function and the symmetry set appears as $A_1 A_1$ or $A_1 A_2$ or $A_1 A_1 A_1$. The $A_2$-set

**Fig. 7.13.** Self-intersection of $\mathscr{D}_{\tilde{F}}$

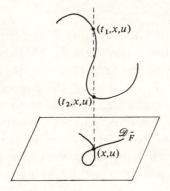

**Fig. 7.14.**

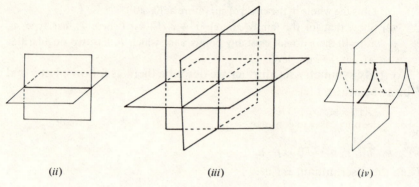

(*ii*)                    (*iii*)                    (*iv*)

**Fig. 7.15.** Local structure of the symmetry set

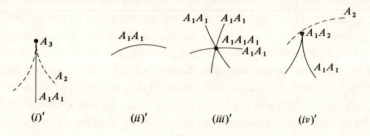

(*i*)'                    (*ii*)'                    (*iii*)'                    (*iv*)'

**Fig. 7.16.**

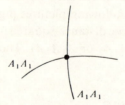

(regular part of the evolute) is drawn as a dotted line in (*i*)' and (*iv*)'. Also, two configurations (*ii*) may project to intersecting lines (fig. 7.16) where the intersection point represents two $A_1$s on one level $F(t_1, x) = F(t_2, x) = u_1$ and two others at another level $F(t_3, x) = F(t_4, x) = u_2$. This happens with the ellipse. All of these are stable, i.e. unchanged by small perturbations of $\gamma$. It is interesting to note that three *concurrent* curves as in (*iii*)' can be stable in this way: so long as they come from three $A_1$ singularities at the same level (i.e. a circle touching the curve in three places), a small perturbation of $\gamma$ will not separate them into three lines forming a small triangle. However *four* concurrent lines in the symmetry set would separate under perturbations of $\gamma$.

## 7.17    Exercises

(1) Show that the conditions for $\tilde{F}_{x,u}$ to have type $A_k$ at $t_0$ are identical with those for $F_x$, and that $\tilde{F}$ is automatically a versal unfolding when $k \leqslant 3$. (The calculations are more or less the same as those in 7.1.) The same remark as 7.3 on genericity now shows that, for general curves, these are the only cases to consider.

(2) Convince yourself (or even better someone else) that the possibilities listed above for the local structure of the symmetry set are the only stable ones. For example, why does one not expect a cusp on the symmetry set to lie also on another branch (the right-hand diagram of fig. 7.8)?

(3) What would four concurrent curves in the symmetry set separate into under perturbations of $\gamma$? Is there more than one possibility for the arrangement of $A_1$ singularities which could (unstably) give four concurrent curves?

(4) What is the geometrical significance of $A_1 A_1$ singularities of *height* functions on a plane curve? What do you expect the local structure of this symmetry set to be? Now try the case of height functions on a *space* curve. The calculations are similar to those above for distance-squared functions on a plane curve.

## Caustics by reflexion

Consider rays of light emanating from a point source $S$ in the plane, reflected from a smooth curve $M$ which we refer to as a *mirror*. We shall always assume that $S$ is not a point of $M$. In 5.33(2) it is shown (with some details to be filled in) that the caustic by reflexion of $M$ relative to $S$, that is the envelope of reflected rays, is also the envelope of normal lines of the orthotomic of $M$ relative to $S$, i.e. the evolute of the orthotomic. See the upper right diagram of fig. 5.1, and fig. 5.15. The orthotomic relative to $S$ is the locus of reflexions of $S$ in the tangent lines to $M$, and we shall study the evolute of $W$ directly from $M$, by a method which involves the *contact of M with conics*. (For further details, see Bruce, Giblin and Gibson (1981).) *We assume below that M has no inflexions; then W is a regular curve by* 5.33(2).

Consider $P \in M$ and suppose that $S$ does not lie on the tangent to $M$ at $P$ (see 7.19 for the contrary case). Let $Q$ be the corresponding point of $W$ and suppose first that $Q$ is not an inflexion of $W$. Then the circle of curvature of $W$ at $Q$ has its centre $F$ on the line $QP$ which is normal to $W$ at $Q$. This point $F$ is the point of the evolute of $W$ corresponding to $Q \in W$, i.e. the point of the caustic corresponding to $P \in M$. It turns out (exercise

7.21(5) below) that the (unique) curve whose orthotomic relative to $S$ is this circle is a conic passing through $P$ with $S$ and $F$ as its foci, an ellipse if $S$ lies inside the circle and a hyperbola if $S$ lies outside (and $S$ cannot lie on the circle) (fig. 7.17).

Now the order of contact of $M$ and the conic at $P$ will be *the same* as the order of contact of their orthotomics, $W$ and the circle, at $Q$. There are several ways of proving this: one way is to use the corresponding result for duals (5.33(9)) and the identification of duals and orthotomics (5.32). Another way, at any rate for small orders of contact, is to use the criterion 4.27(4) for $k$-point contact of two curves ($k \geqslant 3$): they should have the same tangent, curvature and first ($k-3$) derivatives of curvature with respect to arclength at the point. This method is used in the paper cited above, and some start is made on it in 7.21(14) below. The first method is given, in a more general context, in Bruce, Giblin and Gibson (1982).

When $Q$ is an inflexion of $W$ the circle becomes the tangent line at $Q$, the conic becomes a parabola, $F$ recedes to infinity and the same result on contact holds.

Now drawing on our knowledge of evolutes (7.2, 7.11) and contact between curves and circles or lines (2.10, 2.11) we find the following.

### 7.18    Proposition

*Assume $S$ is not on the tangent line to $M$ at $P$. Then*

(i) *The conic constructed above – focus $S$, having orthotomic relative to $S$ equal to the circle of curvature of $W$ at $Q$ or to the tangent line at $Q$ if $Q$ is an inflexion – has at least 3-point contact with $M$ at $P$.*

(ii) *Its focus other than $S$ is the point of the caustic corresponding to $P \in M$ (when the conic is a parabola the focus is 'at infinity' on the reflected*

**Fig. 7.17.** Conic tangent to the mirror and circle of curvature of the orthotomic

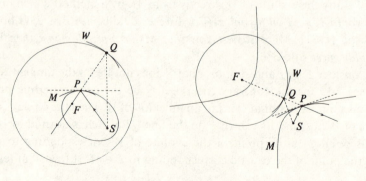

*ray QP). M and the conic have the same tangent and curvature at P (by 4.27(4), or directly by 7.21(14) below).*

(iii) *If the conic has exactly 3-point contact then near F on the caustic, when this is a finite point, the caustic is a smooth (regular) curve.*

(iv) *If the conic and M (or the orthotomic W and the circle or line) have at least 4-point contact, then the tangent, curvature and first derivative with respect to arc length agree for M and the conic at P (or for W and the circle or line at Q).*

(v) *If the contact is exactly 4-point, and F is a finite point, then W has an ordinary vertex at Q and the caustic has an ordinary cusp at F.* □

**7.19** **Proposition** (See exercise 7.21(9).)

*When S lies on the tangent to M at P, the following holds.*

(i) *Q = S, and the radius of curvature of W at Q is equal to the distance QP, so that the point of the caustic corresponding to P is P itself* (fig. 7.18).

(ii) *The caustic is always a regular curve near P and touches M there.* □

Notice that the conic construction breaks down completely in this case: in fact the conic degenerates into the tangent line to $M$ at $P$.

Let us fix a point $P \in M$ and ask: what positions of $S$ will make the caustic singular (non-regular) at the point corresponding to $P$? By 7.18 and 7.19 we look at conics with one focus $S$ and having at least 4-point contact with $M$ at $P$ (fig. 7.19; the other focus will then be the corresponding point of the caustic, or will have receded to infinity). Take $M$ to be parametrized $\gamma(t) = (t, Y(t))$ where $P = (0, 0)$, $Y(0) = Y'(0) = 0$ and $Y(t) = a_2 t^2 + a_3 t^3 + t^4 Y_1(t)$ ($a_2 \neq 0$ since $M$ has non-vanishing curvature), as in 3.4. Write $S = (u, v)$ and $F = (-\lambda u, \lambda v)$ (recall $F$ is on the reflected ray), where $w \neq 0$ since, by 7.19, $S$ does not lie on the tangent to $M$ at $P$.

Any conic touching the $x$-axis at $(0, 0)$ has the form

$$ax^2 + 2hxy + by^2 + y = 0$$

for constants $a$, $h$, $b$. Straightforward calculations (compare for example Sommerville (1924), p. 140) show that $(u, v)$ is a focus of this conic iff

$$4(h^2 - ab)(v^2 - u^2) = 4av + 4hu + 1 \quad \text{and} \quad 2(h^2 - ab)uv = au - hv.$$

**Fig. 7.18.** Incident light tangent to the mirror

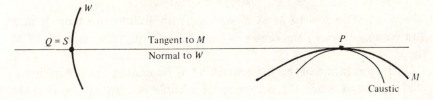

**Fig. 7.19.** $M$ locally as the graph of a function at $(0, 0)$

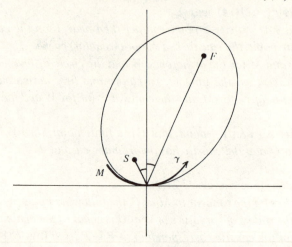

Substituting $x = t$, $y = Y(t)$ in the left-hand side of the equation of the conic we find that they have at least 4-point contact iff $a + a_2 = 0$ and $2ha_2 + a_3 = 0$. However the above conditions on $(u, v)$ give

$$2(v^2 - u^2)(au - hv) = uv(4av + 4hu + 1).$$

Putting $a = -a_2$, $h = -a_3/2a_2$ gives the following:

**7.20**    *The locus of foci $(u, v)$ of conics having at least 4-point contact with $M$ at $(0, 0)$, hence the locus of positions for $S = (u, v)$ for which the caustic is non-regular at the point corresponding to $P$, is the cubic curve*

$$(u^2 + v^2)(2a_2^2 u + a_3 v) = uva_2. \qquad \square$$

Notice that this locus will automatically contain all the positions of $F$ as well as those of $S$ giving non-regularity at $F$. Notice also that the cubic is a *nodal* curve, with node at $P = (0, 0)$ and tangents at the node along the $x$ and $y$ axes (fig. 7.20, left).

It is easy to check that the cubic curve is irreducible (does not contain a line component) iff $a_3 \neq 0$, i.e. iff $P$ is not a *vertex* of $M$ (see 2.15(1)). When $a_3 = 0$ the cubic factorizes as

$$u\left(u^2 + \left(v - \frac{1}{4a_2}\right)^2 - \frac{1}{16a_2^2}\right) = 0,$$

which is the normal to $M$ at $P$ together with a circle touching $M$ at $P$ and passing through the centre of curvature $C = (0, 1/2a_2$ at $P$ (fig. 7.20, right).

As an application of this last remark let $M$ be a circle, centre $C$. Then $a_3$ is always zero, since the curvature of a circle is constant, so that the

**Fig. 7.20.** The nodal cubic of (7.20)

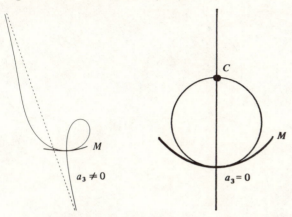

$a_3 \neq 0$

$a_3 = 0$

nodal cubic curve always breaks up into the diameter through $P$ and the circle on $PC$ as diameter. Given $S$ inside $M$, and not at $C$, it is easy to see that there are four positions for $P$ such that the cubic (circle + line) passes through $S$. They are marked $P_1$, $P_2$, $P_3$, $P_4$ in fig. 7.21. (The circles with diameters $P_3C$ and $P_4C$ are suppressed.) Since a conic other than $M$ itself cannot have more than 4-point contact with $M$ (two curves of degree 2 meet in at most 4 points altogether) this shows that *the caustic has exactly four cusps, all ordinary.* That corresponding to $P_i$, $i = 1, 2$ is at the reflexion of $S$ in $P_iC$. For the others see 7.21(3). Notice that the same remark about contact holds when $M$ is any conic, so the caustic of any conic has singularities which are all ordinary cusps. When $M$ is an ellipse it turns out that there are 4 cusps for $S$ inside $M$ and 2 for $S$ outside. Figure 7.22 shows the caustic by reflexion of the circle $x^2 + y^2 = 1$ with the source $S$ at various positions, and likewise some caustics by reflexion of the ellipse $\frac{1}{4}x^2 + y^2 = 1$.

**Fig. 7.21.** A circular mirror, $S$ inside

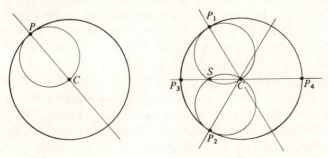

**Fig. 7.22.** Some caustics by reflexion of a circle and an ellipse

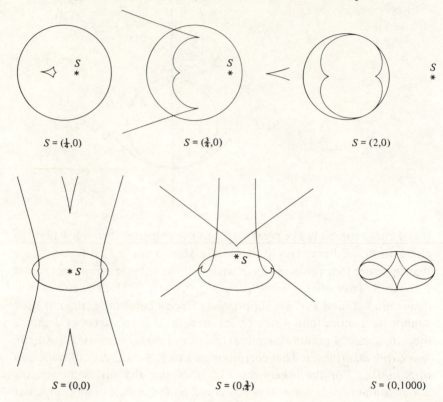

$S = (\tfrac{1}{4},0)$            $S = (\tfrac{3}{4},0)$            $S = (2,0)$

$S = (0,0)$            $S = (0,\tfrac{3}{4})$            $S = (0,1000)$

## 7.21     Exercises

In (1)–(4) suppose that $S$ lies on the normal to $M$ at $P$ (but $S \neq P$). Then $PS$ is also the normal to the conic (as in 7.18) at $P$, so $PS$ is the axis of this conic containing the foci $S$ and $F$ (for a parabola, $F$ is at infinity on the axis). Also $M$ and the conic will have the same centre of curvature $C$ at $P$.

(1) Suppose that the conic of 7.18 is an ellipse so that $S$, $F$ are the foci (but maybe $S$ is the further focus from $P$). $C$ will lie somewhere between $S$ and $F$. Writing $a$, $b$ for the semi-axes and $e$ for the eccentricity of the ellipse $(b^2 = a^2(1 - e^2), 0 \leq e < 1, a \geq b > 0)$, then $SC = CF = ae$, $PC = a$, $PS = a(1 \pm e)$ $= r$, say. Further $\rho = $ radius of curvature at $P = b^2/a$ (see 2.26(1)). Show that $e = \pm (r - \rho)/r$ and, provided $2r > \rho$, $a = r^2/(2r - \rho)$ and the distance $PF$ is $a(1 \mp e) = \rho r/(2r - \rho)$.

(2) By considerations such as (1), show that
 (i) If $S$ and $C$ are on the same side of $P$ and $2r > \rho$ then $F$ is on the same side of $P$, at a distance $\rho r/(2r - \rho)$, and the conic of 7.18 is an ellipse.
 (ii) If $S$ and $C$ are on the same side of $P$ and $2r < \rho$ then $F$ is on the opposite side of $P$, at distance $\rho r/(\rho - 2r)$, and the conic is a hyperbola.

(*iii*) If $S$ and $C$ are on opposite sides of $P$, then $F$ is on the same side of $P$ as $C$, at a distance $\rho r/(2r+\rho)$ from $P$, and the conic is a hyperbola.

(*iv*) With (*i*) or (*ii*) if $2r=\rho$ then the conic is a parabola and $F$ is at infinity on the normal at $P$.

(3) Armed with (2) calculate the positions of the cusps on the caustic of a circle $M$ corresponding to $P=P_3$ and $P=P_4$ in fig. 7.21. (Take also $S$ outside $M$, with these $P_i$ on the diameter through $S$.)

(4) Still with $S$ on the normal at $P$, use the fact that the conic must touch $M$ at a vertex of the conic to deduce that there is at least 4-point contact (hence a cusp or worse on the caustic) iff $P$ is a vertex of $M$.

(5) Show that the orthotomic of an ellipse or hyperbola relative to one focus is a circle centre the other focus. Show that the orthotomic of a parabola relative to its focus is the directrix (i.e. the line perpendicular to the axis, through the reflexion of the focus in the vertex). (Hint. Taking an ellipse in the form $x^2/a^2+y^2/b^2=1$, it is not hard to verify by coordinate geometry that the distance $SQ$, where $S=(ae, 0)$ is one focus and $Q$ is the reflexion of the other focus $(-ae, 0)$ in the tangent line to the ellipse at $(a\cos t, b\sin t)$, equals $2a$ for all $t$. So $Q$ moves on a circle centre $S$. Using geometrical properties of the ellipse the proof is even easier.) Compare 7.14(2).

(6) Take $S$ at the origin 0 (not necessarily on the normal to $M$ at $P$) and parametrize $M$ (near $P$) by $\gamma: I \to \mathbb{R}^2$ (unit speed). Suppose $\gamma(t)$ is never 0, and $\kappa(t)>0$ for all $t$; thus the orthotomic $\delta$ (relative to 0) is a regular curve by 5.33(2). Let $p(t)=-\gamma(t)\cdot N(t)=\pm$ the perpendicular distance from 0 to the tangent to $M$ at $P=\gamma(t)$. (The sign is positive iff $SP$ turns anticlockwise as $t$ increases.) Let $r=\|\gamma(t)\|$. Prove that the curvature $\kappa_0$ of the orthotomic at $Q=\delta(t)$ is

$$\kappa_0 = \frac{2r^2\kappa - p}{2r^3\kappa}.$$

(Without a trick this is rather tedious to verify. The trick is to use 'pedal coordinates', in fact to parametrize $M$ (near $P$) by the distance $r$ from 0. This is valid, near $P$, provided $r'$ does not vanish at $P$, i.e. so long as $S=0$ is *not* on the normal to $M$ at $P$ (but see below!). Assume also that $p \neq 0$, so $S$ is not on the tangent to $M$ at $P$ either. A well known formula for the curvature says that $\kappa = (1/r)(dp/dr)$: this is easy to check using $N' = -\kappa T$, etc., or you can look it up in most books on differential geometry of plane curves. Now if $p_0$, $r_0$ refer to the orthotomic then it is immediate from similar triangles that $p_0 = 2p^2/r$, while clearly $r_0 = \pm 2p$ (fig. 7.23). Using $\kappa_0 = (1/r_0)dp_0/dr_0$ now gives the required formula. (Notice that $OQ$ turns anticlockwise as $t$ increases when $\kappa > 0$; this is implied by the fact that $QP$ is the *oriented* normal to $W$ at $Q$ – see 5.33(2).) The restriction that $S$ should lie off the tangent and normal to $M$ at $P$ can now be removed, for all the functions involved are continuous, so the result must hold in these limiting cases – so long as $S$ is not at $P$, of course.)

**Fig. 7.23.** Pedal coordinates on $M$ and the orthotomic

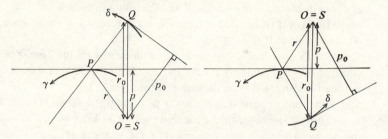

(7) Use the formula in (6) to check the formulas of (2). (Recall $F$ is the centre of curvature of $W$ at $Q$. Notice $p = \pm r$ in the situation of (2).)

(8) Use the formula of (6) to prove that the derivative of $\kappa_0$ with respect to arc length on $W$ is

$$\frac{-3r^2 r' \kappa^2 + 3pr'\kappa + pr\kappa'}{4r^5 \kappa^3}.$$

(First show $(r')^2 = (r^2 - p^2)/r^2$ and $p' = \kappa r r'$. In fact $\kappa_0' = d\kappa_0/dt$ is $2r\kappa$ times the above formula. However $\delta = 2(\gamma \cdot N)N$ shows that $\|\delta'\| = 2\kappa \|\gamma'\| = 2\kappa r$. Again this formula is valid for any position of $S$ except $P$.)

(9) Use (6) and (8) to check 7.19. (In this case $p = 0$ and $r' \neq 0$ (since $rr' = \gamma \cdot T$). You want to find $\kappa_0$ at $Q$ and to check that $\kappa_0' \neq 0$ there.) Also use (6) and (8) to check (4) above.

(10) Let $M$ be the ellipse $x^2/a^2 + y^2 = 1$ and take $S$ on the minor axis $x = 0$, say $S = (0, v)$. Show that when $v > 1$ the point $F_1$ (necessarily an ordinary cusp) on the caustic corresponding to $(0, 1)$ on $M$ is at $(0, (2a^2 - a^2 v + 2v - 2)/(a^2 + 2v - 2))$. What point does this approach as $v \to \infty$?

(11) With $M$, $S$ as in (10), find the position of the point $F_2$ on the caustic corresponding to $(0, -1)$ on $M$. You will need to split into cases according as $v$ is $<$, $=$ or $> \frac{1}{2}a^2 - 1$. What position does $F_2$ approach as $v \to \infty$? Find a value of $a$ for which the limiting position of $F_2$ and that of $F_1$ (see (10)) coincide.

(12) In (9), find a value of $a$ for which the limiting position of $F_1$ (as $v \to \infty$) as in (10) is at $(0, -1)$. Where then is the limiting position of $F_2$ in (11)?

(13) Deduce what you can about the direction in which the cusps on the caustic point when $M$ is a circle. Concentrate on the cusps corresponding to positions of $P \in M$ on the diameter through $S$. (Compare 7.4(1).)

(14) Use the formulas for $\kappa_0$ and its derivative in (6) and (8) to check that two curves have $k$-point contact at $P$ (not an inflexion of either curve) iff their orthotomics (relative to the same point $S$) have $k$-point contact at the corresponding point $Q$, in the cases $k = 2$, 3 and $\geq 4$. (Compare 4.27(4).)

(15) Prove that at a point of $M$ there is a *unique* conic having at least 5-point contact with $M$ (remember that $M$ has non-vanishing curvature). Deduce that the only positions for $S$ which make $P \in M$ give a higher cusp on the

caustic (a singularity other than an ordinary cusp) are the foci of this conic. As $P$ moves round $M$ these foci will trace out a curve called the *focoid* of $M$ (see Bruce, Giblin and Gibson 1981, 1982), and for $S$ off this curve there will be no higher cusps on the caustic: positions of $S$ giving higher cusps are *exceptional*. In fact the focoid is part of the envelope of the nodal cubics in 7.20, the rest of this envelope being $M$ itself, but this is tricky to prove.

## Duals and differential equations

Let $F: \mathbb{R}^3 \to \mathbb{R}$ be smooth and consider the first order differential equation $F(\mathrm{d}y/\mathrm{d}x, x, y) = 0$. We use coordinates $p$, $x$, $y$ in $\mathbb{R}^3$ and shall assume that 0 is a regular value of $F$, so that $M = F^{-1}(0)$ is, in a neighbourhood of any of its points, a parametrized surface in $\mathbb{R}^3$.

If $\partial F/\partial p \neq 0$ at some point $(p_0, x_0, y_0)$ of $M$ then, by the Implicit Function Theorem, we can locally write $p$ as a function $f(x, y)$, so that the differential equation becomes $\mathrm{d}y/\mathrm{d}x = f(x, y)$. This form is far more convenient, since we can obtain series solutions by repeated differentiation (when $F$, and hence $f$, is analytic), while the approximate formula $\delta y \doteq f(x, y)\delta x$ will yield approximate solution curves (called Euler curves).

Geometrically the conditions $F = \partial F/\partial p = 0$ arise when projecting the surface to the $(x, y)$-plane in the $p$-direction; the image of the locus $F = \partial F/\partial p = 0$ under this projection is the apparent contour of $M$ in the $p$-direction. Compare 5.4(3) and 7.13; here $p$ replaces $t$ for reasons of tradition. It is of interest, then, to describe the solution curves of the differential equation given by $F$ when they pass through points of the apparent contour of $M$. We call these curves *solution curves of $F$*.

### 7.22    Examples

(1) $F(p, x, y) = p^2 - 4y$. The apparent contour is the $x$-axis in the $(x, y)$-plane and the equation $(y')^2 = 4y$ has solutions $y = (x - c)^2$ where $c$ is constant. These are parabolas touching the $x$-axis (fig. 7.24, left half).

**Fig. 7.24.** Solution curves of $F$ in 7.22

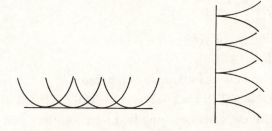
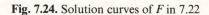

Compare 5.38; the envelope $y=0$ is called a *singular solution* of the equation.

(2) $F(p, x, y)=p^2-\frac{9}{4}x$. The apparent contour is the $y$-axis and the solutions are $(y-c)^2=x^3$ (Fig. 7.24, right half.) In this case $x=0$, the envelope, is *not* a solution.

We shall see later that one of these types of behaviour is more 'typical' than the other.

Consider a (regular) plane curve $\gamma(X)=(X, Y)$ where $Y$ is a smooth function of $X$. The tangent line at $(X, Y)$ is $y=Y'x-(XY'-Y)$, and the curve $\delta(X)=(Y', XY'-Y)$ can be regarded as the dual curve to $\gamma$ (compare 5.32, 5.33(7)). It is easy to check that, so long as $Y''(X)\neq0$, the gradient of the curve $\delta$ at the point with parameter $X$ is precisely $X$.

Now let $L: \mathbb{R}^3\to\mathbb{R}^3$ be the global diffeomorphism (*Legendre transformation*)

$$L(p, x, y)=(x, p, xp-y)=(P, X, Y),$$

which satisfies $L^{-1}=L$, and let $F: \mathbb{R}^3\to\mathbb{R}$ be a smooth function giving a differential equation as at the beginning of this section. Then $G=F\circ L$ gives another differential equation, and we have:

### 7.23    Lemma

*If $\gamma$ is a solution curve of $G$ then $\delta$ is a solution curve of $F$, at any rate at points where $Y''(X)\neq0$.*

**Proof** $0=G(Y', X, Y)=F(X, Y', XY'-Y)$, and $\delta$ has slope $X$.    $\square$

Note that $Y''=0$ is the condition for the solution curve of $G$ to have an inflexion at the point with parameter $X$ (see 2.12(2)) and this will produce a cusp, or worse, on the dual (see 7.7). We usually still think of these bad points as part of the solution curve of $F$.

### 7.24    Example

Let $F(p, x, y)=p^3+yp+x$. Then

$$G(P, X, Y)=F(X, P, XP-Y)=P(1+X^2)+X^3-XY$$

so the new equation is

$$\frac{dY}{dX}-\frac{XY}{1+X^2}=-\frac{X^3}{1+X^2}.$$

Solving by an integrating factor gives

$$Y=-(2+X^2)+c(1+X^2)^{\frac{1}{2}},$$

and the original equation now has solutions of the form

$$x=-2X+cX(1+X^2)^{-\frac{1}{2}}, y=2-X^2-c(1+X^2)^{-\frac{1}{2}}.$$

Try sketching some of these solutions; one is shown together with the apparent contour in fig. 7.25.

**Fig. 7.25.** Apparent contour and one solution curve for $F$ in 7.24

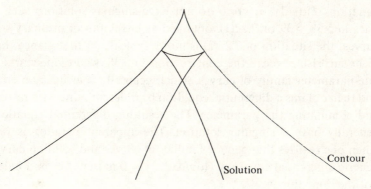

Note that in the example $\partial G/\partial P = 1 + X^2 \neq 0$, whereas $\partial F/\partial p$ is zero on the curve $27x^2 + 4y^3 = 0$, the apparent contour of $F = 0$ in the $p$-direction. Thus $G$ gives a 'better' differential equation than $F$. In fact whenever $G = F \circ L$ we have

$$\frac{\partial G}{\partial P} \circ L = \frac{\partial F}{\partial x} + p \frac{\partial F}{\partial y}.$$

Thus $\partial G/\partial P$ is nonzero at $(P, X, Y) = L(p, x, y)$ if and only if $\partial F/\partial x + p \partial F/\partial y$ is nonzero at $(p, x, y)$. When, as in the example, this always holds, we can (in principle) write $G(\mathrm{d}Y/\mathrm{d}X, X, Y) = 0$ locally in the form $\mathrm{d}Y/\mathrm{d}X = g(X, Y)$, solve, and dualize to obtain parametrized solutions curves of $F$.

Let $\gamma(X) = (X, Y)$ as above be a solution curve of $G$, defined say for $X$ near $X_0$. Suppose also that $\partial F/\partial x + p \partial F/\partial y$ is never zero.

### 7.25 Proposition

*The curve $\gamma$ has an inflexion at $(X, Y)$ if and only if $(p, x, y)$ $= (X, Y', XY' - Y)$ lies on the fold curve $\partial F/\partial p = 0$. Further the inflexion is ordinary if and only if $\partial^2 F/\partial p^2$ is nonzero at $(p, x, y)$.*

Note that the latter condition says that $(p, x, y)$ is an ordinary fold point of the projection $F^{-1}(0) \to \mathbb{R}^2$ given by $(p, x, y) \mapsto (x, y)$ (compare 7.13 and 5.21). When this happens the dual curve $\delta$ will have an ordinary cusp at $(x, y)$ (see 7.7), so the solution curve of $F$ will have an ordinary cusp. In 7.24 this happens at all points of the apparent contour except for $x = y = 0$. See fig. 7.25.

*Proof* This is a matter of differentiating the identity

$$F(X, Y', XY' - Y) = 0$$

with respect to $X$ and using the fact that, at $(p, x, y)$, $\partial F/\partial x + p \partial F/\partial y$ is nonzero. $\qquad\square$

The above reasoning suggests that generally speaking differential equations $F(dy/dx, x, y)=0$ do not have singular solutions such as the ones in 5.38, 5.39 or 7.22(1), obtained as envelopes of ordinary solution curves; the situation of 7.22(2) is more 'typical'. At first glance this may seem surprising: recall the construction in 5.38, where one starts with a one-parameter family of curves, which generally *does* have an envelope, and then obtains a differential equation by differentiating with respect to $x$ and eliminating the parameter. The resulting differential equation *does* generally have a singular solution. The apparent paradox is resolved when one realizes that generic families of curves and solution curves of a generic differential equation $F(dy/dx, x, y)=0$ need not be and indeed are not the same things.

**7.26    Exercise**

Consider a *family* of solution curves $\gamma_c$ of $G$, corresponding to different initial conditions, say $Y(X_0)=Y_0+c$, with $c$ close to 0. Writing $Y(x, c)$ for the corresponding solution ($X$ close to $X_0$) we shall have $\partial Y/\partial c(X_0, 0)\neq 0$. We can examine the duals of all these curves (the solution curves of $F$, in the notation of 7.23), by means of a family of height functions (compare 7.7):

$$\tilde{H}: \mathbb{R} \times S^1 \times \mathbb{R}^2 \rightarrowtail \mathbb{R}$$

where

$$\tilde{H}(X, u, c, v)=(X, Y(X, c))\cdot u-v.$$

The section $c=$constant of $\mathscr{D}_{\tilde{H}}$ is the dual of the solution curve $\gamma_c$. The normal to $\gamma_0$ at $X_0$ is in the direction $(-Y'(X_0, 0), 1)$ so we can examine the map $\tilde{H}$ locally by using instead

$$F: \mathbb{R} \times \mathbb{R}^3, (X_0, 0)\rightarrow \mathbb{R}$$

given by

$$F(X, \lambda, c, v)=(X, Y(X, c))\cdot(\lambda-Y'(X_0, 0), 1)-v,$$

the variable $\lambda$ replacing the point of $S^1$ and giving varying directions. The sections of $\mathscr{D}_F$ will still be locally diffeomorphic to the duals of the curves $\gamma_c$. Let $L(Y'(X_0, 0), X_0, Y_0)=(p_0, x_0, y_0)$.

(i) Show that $F_0$ has type $A_k$ at $X_0$ if and only if $Y^{(i)}(X_0, 0)$ is zero for $i=2,\ldots,k$ and nonzero for $i=k+1$. Show further that this holds if and only if $\partial^i F/\partial p^i(p_0, x_0, y_0)$ is zero for $i=1,\ldots,k-1$ and nonzero for $i=k$.

(ii) Show that $F$ always versally unfolds $F_0$ of type $A_2$ at $X_0$ and that when $F_0$ has type $A_3$ it is versally unfolded provided

$$\frac{\partial^2 F}{\partial y\partial p}\frac{\partial F}{\partial x}\neq\frac{\partial^2 F}{\partial x\partial p}\frac{\partial F}{\partial y}$$

at $(p_0, x_0, y_0)$. (Compare 7.13.)

In the latter case the surface of duals will be locally a swallowtail. However we cannot prove here that the individual duals will in general look like the usual swallowtail sections as in figs. 5.17 and 7.25. This is shown in Bruce (1983), using results of Arnold (1976).

**7.27    Projects**

(1) What information does the method of the present chapter give on the local structure of the envelopes (*i*)–(*iv*) of 5.41?

(2) Those with access to computer graphics will find caustics a rich field for exploration. It is not hard to write a program which will draw the caustic of, say, an ellipse relative to any given point $S$ as source. The caustic can be calculated as the evolute of the orthotomic relative to $S$ (using for example the formula for $\kappa_0$ in 7.21(6) to find the centre of curvature of the orthotomic). It is also possible to work with foci of conics as in 7.18. Try drawing the orthotomic too for some positions of $S$, and also try drawing the reflected rays (normals to the orthotomic) directly, so that the caustic appears as an envelope. For an ellipse as mirror, take $S$ at large distances and see how the cusps move around. With $S$ inside a suitable ellipse the caustic can be made to have four separate components (one cusp on each). Another good curve to take is $y^2 = a^2(x - x^3)$ for various $a > 0$, where $0 \leqslant x \leqslant 1$ so that the curve is a closed loop. Parametrize in two halves, $y \geqslant 0$ and $y \leqslant 0$; a good choice of $x$-values is $x = \frac{1}{2} + \frac{1}{2}\cos t$ for $t$ equally spaced between 0 and $2\pi$, changing from $y = a\sqrt{(x - x^3)}$ to $y = -a\sqrt{(x - x^3)}$ at $t = \pi$. The cubic curves of 7.20 are also good to draw, and the greatest challenge of all is to draw the focoid of 7.21(15)! There is a picture of one for a cubic curve as above ($a = 0.8$) in the first paper cited in 7.21(15).

(3) Plotting symmetry sets of curves by computer graphics is a challenge of a different kind, for the symmetry set is not a parametrized curve – its definition involves 'global' considerations. Try to devise a general method which gives an acceptable approximation for an ellipse (fig. 7.12) and then try your method for a limaçon (see 2.15(5)).

# 8

## *Transversality*

'It has long been an axiom of mine that the
little things are infinitely the most important.'
(*A Case of Identity*)

If one were asked to give Sherlock Holmes a clue as to how he might best
succeed in unlocking the secrets of the smooth Universe he inhabits, one
might do worse than observe that most pairs of lines in the plane meet
once, and most pairs of lines in space do not meet at all (fig. 8.1). Like
Dr Watson you may find it hard to believe that such trivial observations
can form the basis for exploration of what we have assured you is a rich
and beautiful world. But of course all of the best ideas are simple. (And
perhaps one should add, that the most reliable maxims are all clichés.)

The key hinted at above is the idea of *transversality*. Its systematic use,
by René Thom, has proved spectacularly successful in the exploration of
the manifold intricacies of our smooth Universe. As you might imagine
any idea as central and fundamental as transversality (please believe us)
can be approached from many directions. We choose to introduce it as a
generalization of the notion of regular value. The principal technical result
needed to make the whole approach work is indeed nothing more than a
cunningly disguised version of Sard's theorem. This approach may make
some of what follows appear a little technical, with some of the simpler
and more intuitive aspects arising later. The reader is referred to Poston

**Fig. 8.1.** The vital clue

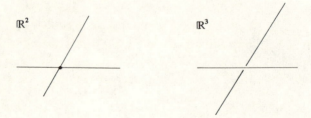

and Stewart (1978) for an alternative approach where increasingly complicated definitions are built up and discussed in some detail culminating in the general one we use here.

In this book we shall use transversality in two ways. Its principal use, in chapter 9, will be to justify the various assertions made in chapter 7 concerning the generic geometry associated to curves. We will also use it to discuss (again) versal unfoldings of smooth functions (see 8.20(5)).

### Smooth manifolds

In this first section we wish to make one final generalization of the idea of a curve. The problem with our current generalization, that of a parametrized manifold, is that it does not include such interesting geometric objects as the unit spheres $S^n \subset \mathbb{R}^{n+1}$ (in particular the circle $x^2 + y^2 = 1$ in the plane). Yet it is evident that in a neighbourhood of any point these spheres *are* parametrized manifolds and should be capable of study by the techniques we have been developing. To overcome this problem we make the following

### 8.1 Definition

A *smooth n-manifold* $X$ is a subset of some Euclidean space $\mathbb{R}^{n+q}$ with the property that each point $x \in X$ has an open neighbourhood $V$ in $\mathbb{R}^{n+q}$ with $V \cap X$ a parametrized $n$-manifold. The notation $X^n$ is often used for a smooth $n$-manifold; $n$ is the *dimension* of $X^n$ and $q$ is the *codimension* of $X$ in $\mathbb{R}^{n+q}$. We sometimes say '$X$ is smooth' to mean '$X$ is a smooth manifold'. Also 'manifold' always means 'smooth manifold'.

### 8.2 Examples

As mentioned above the unit spheres $S^n \subset \mathbb{R}^{n+1}$ are all smooth $n$-manifolds. Indeed it follows from 4.16 that given any smooth map $f: \mathbb{R}^{n+q} \to \mathbb{R}^q$ and any regular value $a \in \mathbb{R}$ the set $f^{-1}(a)$ is a smooth $n$-manifold (or empty of course!). The reader is asked to check that this does follow from 4.16. For the unit spheres one considers as usual $f: \mathbb{R}^{n+1} \to \mathbb{R}$ defined by $f(x) = \|x\|^2$ and the regular value 1. (Note: clearly any parametrized $n$-manifold is automatically a smooth $n$-manifold. The converse is false, the unit circle in $\mathbb{R}^2$ providing a counter-example. See exercise 8.10(1)).

We can now make some further definitions.

### 8.3    Definition

If $X$ is a smooth $n$-manifold, and $x \in X$, the *tangent space to $X$ at $x$* is the tangent space at $x$ to the parametrized manifold $X \cap V$ at $x$ where $V$ is the neighbourhood of 8.1. This tangent space is denoted by $X_x$.

Before proceeding further we need to check that the definition above really makes sense. Looking back at definition 8.1 you will see that we only ask that each point $x \in X$ has a neighbourhood $V$ with $V \cap X$ a parametrized manifold. Thus we have (fig. 8.2) an open neighbourhood $U$ in $\mathbb{R}^{n+q}$ of some point $w \in \mathbb{R}^n \times \{0\}$ (identified with $\mathbb{R}^n$) and a smooth map $\phi: U \to \mathbb{R}^{n+q}$ with $\phi: U \to \phi(U)$ a diffeomorphism and $x = \phi(w)$. Moreover $\phi(U)$ must contain $V$ and $X \cap V$ must coincide with $\gamma(W) \cap V$ where as usual $\gamma$ denotes the restriction of $\phi$ to $\mathbb{R}^n \times \{0\}$ (fig. 8.2). The tangent space to $X$ at $x$ is $T\gamma(w)(\mathbb{R}^n_w)$. The definition 8.3 does *not* however assert that $\phi$ is in any way unique. Indeed thinking of the case of a plane curve, one can *always* reparametrize.

### 8.4    Proposition

*Given any two choices $\phi_1$, $\phi_2$: $\mathbb{R}^{n+q}$, $0 \to \mathbb{R}^{n+q}$ for $\phi$ as above (where we take $w = 0$), let $\gamma_1, \gamma_2$ be the corresponding restrictions to $\mathbb{R}^n \times \{0\}$. Then $T\gamma_1(0)(\mathbb{R}^n_0) = T\gamma_2(0)(\mathbb{R}^n_0)$, i.e. $X_x$ is well defined.*

**Proof** Clearly the maps $\phi_1^{-1} \circ \gamma_2$ and $\phi_2^{-1} \circ \gamma_1$ give local diffeomorphisms $\mathbb{R}^n$, $0 \to \mathbb{R}^n \times \{0\}$, $0$. So $D(\phi_2^{-1} \circ \gamma_1)(0) = (D\phi_2(0))^{-1} \circ D\gamma_1(0): \mathbb{R}^n \to \mathbb{R}^n$ is an isomorphism. Consequently $D\gamma_2(0)(\mathbb{R}^n) = D\phi_2(0)(\mathbb{R}^n \times \{0\}) = D\gamma_1(0)(\mathbb{R}^n)$, as required.    □

We can now define smooth maps between smooth manifolds.

### 8.5    Definition

(*i*) Given smooth manifolds $X^n \subset \mathbb{R}^{n+q}$, $Y^m \subset \mathbb{R}^{m+p}$ a map $f: X \to Y$ is *smooth* if for every $x \in X$ there is an open neighbourhood $U$ of $x$ in $\mathbb{R}^{n+q}$ and a smooth map $F: U \to \mathbb{R}^{m+p}$ with $F|X \cap U = f$.

(*ii*) Given $f$ as above the *tangent map* $Tf(x): X_x \to Y_{f(x)}$ is defined by $Tf(x)(u_x) = DF(x)(u)_{f(x)}$.

Again we have something to check.

**Fig. 8.2.** Smooth manifold

## 8.6      Proposition

*The tangent map $Tf(x)$ is independent of the choice of F.*

**Proof.** Given two parametrizations $\gamma_1\colon \mathbb{R}^n, 0 \to X, x$; $\gamma_2\colon \mathbb{R}^m, 0 \to Y, f(x)$ of neighbourhoods of $x \in X$ and $f(x) \in Y$ respectively, we have a commutative diagram of smooth mappings

$$\mathbb{R}^{n+q}, x \xrightarrow{\;F\;} \mathbb{R}^{m+p}, f(x)$$
$$\gamma_1 \uparrow \qquad\qquad\qquad \uparrow \gamma_2$$
$$\mathbb{R}^n, 0 \xrightarrow[\gamma_2^{-1} \circ f \circ \gamma_1]{} \mathbb{R}^m, 0$$

(where $\gamma_2^{-1} \circ f \circ \gamma_1$ is smooth since it coincides with $\gamma_2^{-1} \circ F \circ \gamma_1$). Taking derivatives we find that $DF(x) \circ D\gamma_1(0) = D\gamma_2(0) \circ D(\gamma_2^{-1} \circ f \circ \gamma_1)(0)$, so clearly $DF(x)$ maps $D\gamma_1(0)(\mathbb{R}^n)$ to $D\gamma_2(0)(\mathbb{R}^n)$. One now easily checks that $Tf(x)$ maps $X_x$ to $Y_{f(x)}$, and since $Tf(x)(u_x) = (D\gamma_2(0) \circ D(\gamma_2^{-1} \circ f \circ \gamma_1)(0) \circ D\gamma_1^{-1}(0)(u))_{f(x)}$, $Tf(x)$ is independent of our choice of $F$.   □

Virtually all of the usual results from calculus have analogues for smooth mappings between smooth manifolds. In particular, just as in 4.22(3) we have the following:

## 8.7      The Chain Rule

*Let $X$, $Y$ and $Z$ be smooth manifolds and $f\colon X \to Y$, $g\colon Y \to Z$ smooth maps. Then $g \circ f\colon X \to Z$ is a smooth map and $T(g \circ f)(x) = Tg(f(x)) \circ Tf(x)$.*

**Proof** Let us suppose $X \subset \mathbb{R}^r$, $Y \subset \mathbb{R}^s$, $Z \subset \mathbb{R}^t$. Since $f$ is smooth for any $x \in X$ there is a smooth map $F\colon \mathbb{R}^r \rightarrowtail \mathbb{R}^s$, defined on a neighbourhood of $x$, with the restriction $F|X$ coinciding with $f$. Similarly we have a smooth map $G\colon \mathbb{R}^s \rightarrowtail \mathbb{R}^t$ defined as a neighbourhood of $f(x)$ with $G|Y = g$. By selecting the neighbourhood $U$ of $x$ in such a way that $F(U) \subset$ domain of $G$ we have a well defined composite $G \circ F\colon \mathbb{R}^r \rightarrowtail \mathbb{R}^t$ which clearly proves that $g \circ f$ is smooth. The identity $T(g \circ f)(x) = Tg(f(x)) \circ Tf(x)$ follows directly from the chain rule for derivatives, since $D(G \circ F)(x) = DG(f(x)) \circ DF(x)$.   □

## 8.8      Definition

If $f\colon X \to Y$ is a smooth map, a point $y \in Y$ is a *regular value* of $f$ if for every $x \in f^{-1}(y)$ the tangent map $Tf(x)\colon X_x \to Y_y$ is surjective.

We can extend the notion of 'set of measure zero' to arbitrary manifolds, and we can then prove the following.

## 8.9      Sard's Theorem (revisited)

*If $f\colon X \to Y$ is a smooth map the set of points of $Y$ which are not regular form a set of measure zero in $Y$. See the appendix.*

**8.10     Exercises**

(1) Show that the unit circle $S^1$ is a smooth manifold in $\mathbb{R}^2$, but is not a smooth parametrized manifold. Define $\theta: \mathbb{R} \to S^1$ by $\theta(t) = (\cos t, \sin t)$. Show that $\gamma: S^1 \to \mathbb{R}^n$ is smooth iff $\gamma \circ \theta: \mathbb{R} \to \mathbb{R}^n$ is smooth. Show also that $\theta$ is smooth.

(2) Let $f: \mathbb{R}^n \to \mathbb{R}^m$ be a smooth map, and define graph $f = \{(x, f(x)): x \in \mathbb{R}^n\}$. Show that graph $f$ is a smooth manifold in $\mathbb{R}^n \times \mathbb{R}^m$ of dimension $n$, and show that the tangent space at $v = (x, f(x))$ is spanned by the $n$ vectors $(0, \ldots, 1, \ldots, 0, \partial f_1/\partial x_i, \ldots, \partial f_m/\partial x_i)_v$ where the 1 is in the $i$th position.

(3) Let $X$ and $Y$ be smooth manifolds. Prove that $X \times Y$ is also a smooth manifold and $(X \times Y)_{(x,y)} = X_x \times Y_y$. Writing $\pi: X \times Y \to Y$ for the projection onto the second factor show that $T\pi(x, y): (X \times Y)_{(x,y)} = X_x \times Y_y \to Y_y$ is also projection onto the second factor.

(4) Prove that the set of $m \times n$ matrices of rank $r$, denoted by $Y_r$, is a manifold (in the space of $m \times n$ matrices $M(m, n) = \mathbb{R}^{mn}$) of dimension $(m+n-r)r$. (Proceed as follows. Suppose first that we are working at an $(m \times n)$ matrix $A$ of the form

$$A = \begin{bmatrix} B & C \\ D & E \end{bmatrix}$$

where $B$ is nonsingular of rank $r$. Postmultiply by the matrix

$$X = \begin{bmatrix} I & -B^{-1}C \\ 0 & I \end{bmatrix}$$

to deduce that $A$ has rank $r$ if and only if $E - DB^{-1}C = 0$. The map

$$\begin{bmatrix} B & C \\ D & E \end{bmatrix} \mapsto \begin{bmatrix} B & C \\ D & E + DB^{-1}C \end{bmatrix}$$

now gives a diffeomorphism $M(m, n) \to M(m, n)$ with the set $U = \{A: \det B \neq 0, E = 0\}$ (an open subset of $\mathbb{R}^{(m+n-r)r}$) mapping to $Y_r$. Deduce the general case from this result.)

(5) Let $X^n$, $Y^m$ be smooth manifolds and $f: X \to Y$ a map. Show that $f$ is smooth if and only if for any $x \in X$, and some parametrizations $\gamma_1$ of $X$ at $x$, $\gamma_2$ of $Y$ at $y = f(x)$ the map $\gamma_2^{-1} \circ f \circ \gamma_1$ is smooth. As usual we shrink the neighbourhood say $U$ of $\gamma_1^{-1}(x)$ in $\mathbb{R}^{n+q}$ so that $f \circ \gamma_1(U)$ is contained in the domain of $\gamma_2^{-1}$.

(Hint. Clearly if $f$ is smooth and $F: \mathbb{R}^{n+q} \to \mathbb{R}^{m+p}$ is a smooth map with $F|X = f$ near $x$, then $\gamma_2^{-1} \circ f \circ \gamma_1 = \gamma_2^{-1} \circ F \circ \gamma_1$ is smooth. Conversely if $\gamma_2^{-1} \circ f \circ \gamma_1$ is smooth and $\pi$ is the natural projection $\mathbb{R}^n \times \mathbb{R}^q \to \mathbb{R}^n$ define $F: \mathbb{R}^{n+q} \to \mathbb{R}^{n+q}$ in a neighbourhood of $x$ by $\phi_2 \circ (\gamma_2^{-1} \circ f \circ \gamma_1) \circ \pi \circ \phi_1^{-1} = F$. Clearly $F|X = f$ and $F$ is smooth.)

(6) Prove the following *Inverse Function Theorem* for maps between smooth manifolds. If $f: X \to Y$ is a smooth map and $Tf(x): X_x \to Y_y$ is invertible then there are neighbourhoods $U$ of $x$ in $X$, $V$ of $y$ in $Y$ with $f: U \to V$ a diffeomorphism, i.e. $f$ is bijective and has a smooth inverse. (Use the result of the previous exercise, and of course the usual Inverse Function Theorem.)

(7) If $f: X^n \to \mathbb{R}^m$ is a smooth map and $y \in \mathbb{R}^m$ is a regular value show that $Y = f^{-1}(y)$ is a smooth manifold, and the tangent space $Y_x$ to $Y$ at $x$ is the kernel of the tangent map $Tf(x)$. (Hint: by 4.17(6) we know that near any point $x \in Y$, $X$ is the inverse image of a regular value 0 of a smooth map $g: \mathbb{R}^{n+q} \to \mathbb{R}^q$. Since $f$ is smooth there is a smooth map $F: \mathbb{R}^{n+q} \to \mathbb{R}^m$ defined on some neighbourhood of $x$ in $\mathbb{R}^{n+q}$ with $F|Y = f$. Show that $(0, y)$ is a regular value of $(g, F): \mathbb{R}^{n+q} \to \mathbb{R}^q \times \mathbb{R}^m$ and $(g, F)^{-1}(0, y)$ and $Y$ coincide near $x$.)

## Transversality

As we have seen in chapter 4 one particularly pleasant method of constructing smooth curves (and manifolds in general) is by taking the inverse images of regular values $y$ of smooth mappings $f: \mathbb{R}^n \to \mathbb{R}^m$. Instead of considering points $y \in \mathbb{R}^m$, suppose we consider some manifold $Y \subset \mathbb{R}^m$. What conditions on $f$ and $Y$, replacing the regular value condition above, would ensure that $f^{-1}(Y)$ is smooth? Let us suppose that $Y$ is itself the inverse image of a regular value $z$ of some smooth map $g: \mathbb{R}^m \to \mathbb{R}^p$. (By 4.17(6) and our definition of smooth manifold this is always true at least in some neighbourhood of any point $y \in Y$.) So $f^{-1}(Y) = f^{-1}(g^{-1}(z)) = (g \circ f)^{-1}(z)$ will be smooth if $z$ is a regular value of $g \circ f$. So for each $x \in f^{-1}(Y)$ we want $D(g \circ f)(x)(\mathbb{R}^n_x) = \mathbb{R}^p_z$. By the chain rule $D(g \circ f)(x) = Dg(f(x)) \circ Df(x)$ and since $z$ is a regular value of $g$ we know that $Dg(f(x)): \mathbb{R}^m_{f(x)} \to \mathbb{R}^p_z$ is surjective. A little thought now shows that $D(g \circ f)(x)(\mathbb{R}^n_x) = \mathbb{R}^p_z$ if and only if $\ker Dg(f(x)) + \operatorname{im} Df(x) = \mathbb{R}^m_{f(x)}$. By 4.24 we know that $\ker Dg(f(x)) = Y_{f(x)}$ so the condition that $z$ should be a regular value of $g \circ f$ can be reinterpreted as

$$\operatorname{im} Df(x) + Y_{f(x)} = \mathbb{R}^m_{f(x)} \text{ for all } x \in f^{-1}(Y).$$

Note that there is now no mention of the mapping $g$. Of course using exercise 8.10(7) we find that the same ideas work for a smooth map $f: X \to \mathbb{R}^m$, with $X^n$ a smooth manifold in some $\mathbb{R}^{n+q}$ and $Y$, as above, a manifold in $\mathbb{R}^m$. Thus $f^{-1}(Y)$ will be a manifold (in $\mathbb{R}^{n+q}$) if

$$\textbf{(T)} \quad \operatorname{im} Tf(x) + Y_{f(x)} = \mathbb{R}^m_{f(x)} \text{ for all } x \in f^{-1}(Y).$$

## 8.11 Definition

If for some smooth map $f: X \to \mathbb{R}^m$ and some manifold $Y \subset \mathbb{R}^m$ the condition **(T)** above holds we say that $f$ is *transverse to* $Y$ and we write $f \pitchfork Y$.

From our work above we immediately deduce the following.

## 8.12     Proposition

*If* $f: X \to \mathbb{R}^m$ *is transverse to* $Y$ *then* $f^{-1}(Y)$ *is a smooth manifold.* □

Note that when $Y$ consists of a single point $y$ the condition (T) that '$f$ should be transverse to $Y$' reduces to '$y$ is a regular value of $f$'. In this sense then transversality *does* generalize the idea of regular value. This analogy is however a little misleading. We shall consider the condition (T) to be a condition on the map $f$ rather than the manifold $Y$. This change of emphasis is very important. Indeed it would be more accurate to state that our main use of transversality of smooth maps $f: X \to \mathbb{R}^m$ to manifolds $Y \subset \mathbb{R}^m$ is as a generalization of $f$ *misses* $Y$, i.e. $f(X) \cap Y = \emptyset$. (This may seem a strange viewpoint but we hope to justify it later on.)

Whichever way one thinks of transversality the idea is a very important one, and provides a tremendously powerful tool for studying smooth maps and manifolds. Now to some examples.

## 8.13     Examples

(1) Define, for any $a \in \mathbb{R}$, $f_a: \mathbb{R} \to \mathbb{R}^2$ by $f_a(x) = (x, a)$, and let $Y \subset \mathbb{R}^2$ be the parabola $\{(x, x^2): x \in \mathbb{R}\}$ (fig. 8.3).

Clearly $f_a(\mathbb{R})$ meets $Y$

(*i*) in 2 points $(\pm a^{\frac{1}{2}}, a)$ if $a > 0$;

(*ii*) in 1 point $(0, 0)$ if $a = 0$;

(*iii*) in 0 points if $a < 0$.

For $a \neq 0$ we claim that the map $f_a$ is transverse to $Y$. For $a < 0$ this is trivially true (if you do not see why turn back to the statement of condition (T)). For $a > 0$ one easily checks that

$$\text{im } Tf_a(\pm a^{\frac{1}{2}}) + Y_{(\pm a^{\frac{1}{2}}, a)} = \text{Sp}\{(1, 0)\} + \text{Sp}\{(1, \pm 2a^{\frac{1}{2}})\} = \mathbb{R}^2_{(\pm a^{\frac{1}{2}}, a)}.$$

Note that transversality here means that $f(\mathbb{R})$ and $Y$ cut cleanly, i.e. are not tangent. This gives the origin of the use of the word 'transverse' in this

**Fig. 8.3**

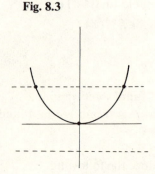

context. Note also that although $f_0$ is not transverse to $Y$ (the above sum of vector spaces is just $\mathrm{Sp}\{(1,0)\}$ when $a=0$), by perturbing it slightly (to $f_a$, any $a \neq 0$ for example) we can make it transverse to $Y$. Indeed we are rather understating our case when we say we can *make $f_0$ transverse to $Y$*. We should have to be very careful when perturbing $f_0$ *not* to ensure transversality.

To further develop our intuition concerning transversality the most illuminating examples to consider are the cases when $f \colon X \to \mathbb{R}^m$ is simply the inclusion map of a manifold $X \subset \mathbb{R}^m$. (Why is this smooth?)

The condition for $f$ to be transverse to a manifold $Y \subset \mathbb{R}^m$ then reduces to
$$X_y + Y_y = \mathbb{R}^m_y \text{ for all } y \in X \cap Y,$$
and if this condition holds we write $X \pitchfork Y$, and say that $X$ *is transverse to $Y$* (or for that matter $Y$ is transverse to $X$, since the condition is clearly symmetric in $X$ and $Y$).

(2) Let $X = \{(x,0) \colon x \in \mathbb{R}\} \subset \mathbb{R}^2$, $Y = \{(0,y) \colon y \in \mathbb{R}\} \subset \mathbb{R}^2$. Clearly $X \pitchfork Y$.

(3) Let $X = \{(x,0,0) \colon x \in \mathbb{R}\} \subset \mathbb{R}^3$, $Y = \{(0,y,0) \colon y \in \mathbb{R}\} \subset \mathbb{R}^3$.
Here $X$ is not transverse to $Y$. For $X$ and $Y$ intersect at the origin $0 = (0,0,0)$ $\in \mathbb{R}^3$ and since dim $X + \dim Y = 2 < 3 = \dim \mathbb{R}^3$ clearly $X_0 + Y_0$ cannot possibly give $\mathbb{R}^3_0$.

(4) Let $X$ be the unit sphere $S^2 = \{x \in \mathbb{R}^3 \colon x_1^2 + x_2^2 + x_3^2 = 1\}$ and $Y$ be the plane $Y_1 = \{(0, x_2, x_3) \colon x_2, x_3 \in \mathbb{R}\}$ (resp. the lines $Y_2 = \{(0,0,x_3) \colon x_3 \in \mathbb{R}\}$, $Y_3 = \{(0, 2, x_3) \colon x_3 \in \mathbb{R}\}$, $Y_4 = \{(x_1, 1, 0) \colon x_1 \in \mathbb{R}\}$; (see fig. 8.4).

In each case other than the last we *do* have transverse intersection. Figure 8.5 gives some further pictorial examples.

Two important points can be gleaned from these examples.

**Fig. 8.4.**

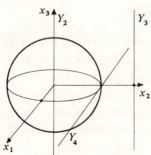

**Fig. 8.5.** Transversality and non-transversality

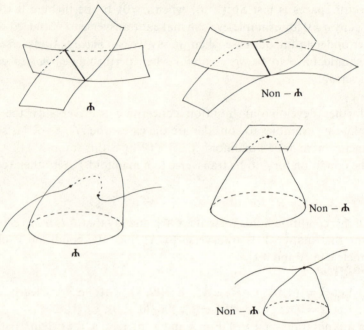

## 8.15     Remarks

(1) *Given $f$: $X \to \mathbb{R}^m$ with $f(X) \cap Y = \varnothing$ then $f$ is transverse to $Y$.* This follows trivially from the definition of transversality, and is the analogue of the remark following 4.12 to the effect that any point not in the image of a smooth map $f$ is a regular value of $f$.

(2) The second, rather more telling, but still trivial observation is as follows. *If $\dim X + \dim Y < m$ then $f$ can only be transverse to $Y$ if $f(X) \cap Y = \varnothing$.*

(***Proof*** $\dim(\operatorname{im} Tf(x)) \leqslant \dim X$ and so if $f(x) \in Y$ for some $x \in X$ we have $\dim (\operatorname{im} Tf(x) + Y_{f(x)}) \leqslant \dim X + \dim Y < m$.)

It is amazing that deep consequences can be derived from such simple observations, but they can, as we shall see. (These remarks partly explain our assertion that transversality generalises '$f$ misses $Y$'.)

One final point. From the above examples one should guess that non-transverse intersection is very rare. In each case arbitrarily small deformation of either manifold produces a transverse intersection of $X$ and $Y$. It turns out that the same is true for smooth mappings $f$: $X \to \mathbb{R}^m$, with

$Y \subset \mathbb{R}^m$: most maps $f$ will be transverse to $Y$. The purpose of the next section is to prove a result which can be used to justify such assertions.

### 8.16    Exercises

(1) Show that $f: \mathbb{R}^n \to \mathbb{R}^m$ is transverse to $Y \subset \mathbb{R}^m$ if and only if graph $f$ and $\mathbb{R}^n \times Y$ are transverse in $\mathbb{R}^n \times \mathbb{R}^m$.

(2) If $\phi: \mathbb{R}^n \to \mathbb{R}^m$ is a linear map and $V \subset \mathbb{R}^m$ is a vector subspace of $\mathbb{R}^m$ show that $\phi \pitchfork V$ if and only if $\mathrm{im}\,\phi + V = \mathbb{R}^m$.

(3) Try to justify the assertion that 'most' linear maps $\phi: \mathbb{R}^n \to \mathbb{R}^m$ are transverse to a subspace $V \subset \mathbb{R}^m$ if and only if $n + \dim V \geq m$.

(4) Which of the following pairs of linear spaces intersect transversally? (Use the criterion developed in problem (2) above, with $\phi$ an inclusion.)

   (*i*)  The $x_1 x_2$-plane and the plane $\mathrm{Sp}\{(0, 1, 1), (1, -1, 0)\}$ in $\mathbb{R}^3$.

   (*ii*)  The plane $\mathrm{Sp}\{(1, 3, 0), (2, -1, 0)\}$ and the $x_1$-axis in $\mathbb{R}^3$.

   (*iii*) $\mathbb{R}^n \times \{0\}$ and the diagonal $\Delta = \{(x, x): x \in \mathbb{R}^n\}$ in $\mathbb{R}^n \times \mathbb{R}^n$.

   (*iv*)  The diagonal $\Delta$ and the skew diagonal $\{(x, -x): x \in \mathbb{R}^n\}$ in $\mathbb{R}^n \times \mathbb{R}^n$.

   (*v*)  The symmetric ($A^T = A$) and skew symmetric ($A^T = -A$) matrices in $M(n, n)$.

   (*vi*)  The symmetric (resp. skew symmetric) matrices and the matrices of trace zero in $M(n, n)$.

(5) If $f: X \to \mathbb{R}^m$ is a smooth map with $f \pitchfork Y$ show that $f^{-1}(Y)_x = Tf(x)^{-1}(Y_{f(x)})$ for any $x \in f^{-1}(Y)$. (Recall the discussion immediately before 8.11.)

(6) Deduce from question (5) above that if $X \pitchfork Y$ in $\mathbb{R}^m$ then $(X \cap Y)_y = X_y \cap Y_y$ for any $y \in X \cap Y$.

(7) If $f: X \to \mathbb{R}^m$ is transverse to $Y$ and $\dim X + \dim Y = m$, show that the points of $f^{-1}(Y)$ are *isolated*. (That is each $x \in f^{-1}(Y)$ has a neighbourhood $U$ with $U \cap f^{-1}(Y) = \{x\}$.) Deduce that, if $X$ is compact, $f^{-1}(Y)$ is finite. (Hint: write $Y$ locally as the inverse image of a regular value of some smooth map $g$, and then apply the Inverse Function Theorem to $g \circ f$.)

## Thom's transversality lemma

In this section we are going to produce a lemma which will allow us to prove results of the form: 'most' maps $f: X \to \mathbb{R}^m$ are transverse to a fixed manifold $Y \subset \mathbb{R}^m$. Since we obtained the notion of transversality by generalizing the idea of regular value one might guess that Sard's Theorem is a key ingredient in the proof. This is indeed the case, but before proving the result we note that we are considering a situation quite different from that of Sard's Theorem. For there we have a fixed map $f: X \to \mathbb{R}^m$ and prove that most points of $\mathbb{R}^m$ are regular values for $f$. The natural generalization of this would be that for most manifolds $Y \subset \mathbb{R}^m$ we have $f \pitchfork Y$. The result we want however is that fixing $Y$ in $\mathbb{R}^m$ most *maps* $f: X \to \mathbb{R}^m$ are transverse to $Y$. The following beautiful result of Thom provides the means of justifying such assertions.

### 8.17     Fundamental Transversality Lemma of Thom

Let $X \subset \mathbb{R}^r$, $Y \subset \mathbb{R}^m$ be smooth manifolds and $U$ an open set in $\mathbb{R}^t$ with $G: X \times U \to \mathbb{R}^m$ a smooth map transverse to $Y$. Then for almost all $a \in U$ (all $a$ outside a set of measure zero) the maps $G_a: X \to \mathbb{R}^m$ given by $G_a(x) = G(x, a)$ are transverse to $Y$.

Before giving the proof we explain why this is the type of result we want. Suppose we are given a smooth map $f: X \to \mathbb{R}^m$ *not* transverse to $Y$. We hope to embed it in a family $G$ as above, i.e. to find a family $G: X \times U \to \mathbb{R}^m$ transverse to $Y$ with say $G_0 = f$. The result then implies that for $a \in U$ arbitrarily close to the origin in $U$ the map $G_a$ is transverse to $Y$, and of course $G_a$ is then $f = G_0$ perturbed slightly. Thus $f$ can be arbitrarily closely approximated by smooth maps $X \to \mathbb{R}^m$ which are transverse to $Y$.

***Proof of Lemma.***     Since $G$ is transverse to $Y$ it follows that $G^{-1}(Y)$ is a smooth manifold in $\mathbb{R}^r \times \mathbb{R}^t$. We claim that $a \in \mathbb{R}^t$ is a regular value of the natural projection $\pi: G^{-1}(Y) \to U$ if and only if $G_a \pitchfork Y$. The Lemma clearly follows from this claim by Sard's Theorem 4.18.

To prove the claim we first reduce to the case when $Y$ is a point. Near each of its points, $Y$ can be written as the inverse image of a regular value, say 0, of some smooth map $g: \mathbb{R}^m \to \mathbb{R}^n$ (compare 4.17(6)). One easily checks that $G$ (resp. $G_a$) is transverse to $Y$ at this point if and only if $g \circ G$ (resp. $g \circ G_a$) has 0 as a regular value. Consequently we may suppose that $Y$ consists of a single point, say $y$.

The condition for $G$ to be transverse to $\{y\}$ is that, for all $(x, a) \in X \times U$ with $G(x, a) = y$, we have

$$TG(x, a)(X_x \times U_a) = \mathbb{R}^m_y. \tag{1}$$

On the other hand $G_a$ is transverse to $\{y\}$ if and only if, for the same $x$s as in (1), we have

$$TG(x, a)(X_x \times \{0\}) = \mathbb{R}^m_y. \tag{2}$$

Finally $a$ is a regular value of the projection $\pi: G^{-1}(y) \to U$ if and only if for the same $x$s as in (1) we have

$$T\pi(x, a)(G^{-1}(y))_{(x,a)} = U_a. \tag{3}$$

that is

$$T\pi(x, a)(\text{kernel } TG(x, a)) = U_a. \tag{4}$$

Using the fact that $T\pi(x, a)$ is the natural projection to $U_a$ (compare 8.10(3)) it is now an easy exercise in linear algebra to show that (1) and (2) hold if and only if (1) and (4) hold.     $\square$

Before moving on to explicit applications we make some remarks about the way this lemma is applied. Usually the map $G$ is constructed so as to be a submersion, and hence transverse to any manifold $Y \subset \mathbb{R}^m$ whatsoever. Thus we need not concern ourselves with $Y$ at all! Even better, one can often construct $G$ in such a way that it is a submersion by virtue of the tangent vectors from the $U$ space of deformations (i.e. $TG(x, a)(\{0\} \times U_a) = \mathbb{R}^m_{G(x,a)}$.) This means we need hardly concern ourselves with $G_0 = f$ either, as we shall see.

Now for an application.

### 8.18    Proposition

*Let $X$ and $Y$ be two smooth manifolds in $\mathbb{R}^n$. Translating $X$ by almost any vector $a \in \mathbb{R}^n$ gives a manifold $X_a$ with $X_a \pitchfork Y$.*

**Proof** Consider the map $G: X \times \mathbb{R}^n \to \mathbb{R}^n$ defined by $G(x, a) = x + a$. Clearly $G$ is a submersion ($TG(x, a)(0, v_a) = v_a$ for any $v_a \in \mathbb{R}^n_a$). Hence $G$ is transverse to $Y$. So by 8.17 for almost all $a \in \mathbb{R}^n$ we have $G_a: X \to \mathbb{R}^n$ transverse to $Y$, hence $G_a(X) = X_a$ transverse to $Y$.     □

Thus if $X$ and $Y$ do not intersect transversally then, by bodily moving $X$ an arbitrarily small amount (choose $a$ close to 0), we can ensure transversal intersection.

Here is another typical application.

### 8.19    Proposition

*Let $f: \mathbb{R}^n \to \mathbb{R}^m$ be any smooth map, and for $A \in M(m, n)$ define $F_A: \mathbb{R}^n \to \mathbb{R}^m$ by $F_A(x) = f(x) + A(x)$. Suppose $m \geqslant 2n$; then for almost all $A \in M(m, n)$ the map $F_A: \mathbb{R}^n \to \mathbb{R}^m$ has maximal rank ($=n$) everywhere, i.e. $F_A$ is an immersion.*

**Proof.** Define $F: \mathbb{R}^n \times M(m, n) \to \mathbb{R}^m$ by $F(x, A) = F_A(x) = f(x) + A(x)$. Consider the associated family $G: \mathbb{R}^n \times M(m, n) \to M(m, n)$ given by $G(x, A) = DF_A(x)$. We claim that $G$ is a submersion. Indeed $DF_A(x) = Df(x) + A$, and so one easily checks that the $(n + mn) \times mn$ Jacobian matrix of $G$ at $(x, A)$ has an $mn$ by $mn$ identity submatrix, and hence has rank $mn$.

Let $Y_r = \{A \in M(m, n): \text{rank } A = r\}$. We have shown that $Y_r$ is a manifold of dimension $(m + n - r)r$ in $\mathbb{R}^{mn} = M(m, n)$ (8.10(4)). Now $G$ is a submersion and hence transverse to each $Y_r$, $0 \leqslant r \leqslant n$. So, by Thom's Lemma, for almost all $A \in M(m, n)$ the map $DF_A = G(-, A): \mathbb{R}^n \to M(m, n)$ is transverse to each of the $Y_r$. (See the note below.) However if $r \leqslant n - 1$ we have $\dim Y_r = (m + n - r)r \leqslant (m + 1)(n - 1)$, so $\dim \mathbb{R}^n + \dim Y_r \leqslant mn - m + 2n - 1 < mn$. Thus in this case $DF_A$ is transverse to $Y_r$ means that $DF_A$ misses

$Y_r$ when $r \leqslant n-1$. Hence $DF_A(\mathbb{R}^n) \subset Y_n$ and the result follows. (Compare 8.15(2).)    □

We can choose $A$ arbitrarily close to the zero matrix in 8.19 and deduce that there exist such $A$(when $m \geqslant 2n$) for which $F_A$ is an immersion. This $F_A$ will closely approximate $f$ on any compact set in $\mathbb{R}^n$. In particular, taking $n=1$ and $m=2$ or 3, it follows that any plane or space curve $\gamma: \mathbb{R} \to \mathbb{R}^m$ can be made regular by adding suitable linear terms to the components of $\gamma$, and the coefficients in these terms can be made arbitrarily small.

Note that in 8.19 we have actually used a slight extension of Thom's Lemma, for we have replaced the single manifold $Y$ by a finite number of manifolds $Y_0, \ldots, Y_{n-1}$ of possibly differing dimensions. The point is of course that we simply use Sard's Theorem in the proof of 8.17 to obtain a bad set of values $a \in U$ of measure zero. One application of Sard's Theorem for each manifold $Y_i$ in $\mathbb{R}^m$ yields a collection of bad sets indexed by $i$. Since the union of countably many sets of measure zero is of measure zero this means that Thom's Lemma remains valid if we replace $Y$ by a countable union of manifolds $\bigcup_{i=1}^{\infty} Y_i$ in $\mathbb{R}^m$.

### 8.20    Exercises

(1) (*i*) Let $U \subset (\mathbb{R}^p)^k$ consist of all linearly independent $k$-tuples of vectors in $\mathbb{R}^p (k \leqslant p)$. Show that $U$ is open.

(*ii*) Define $F: \mathbb{R}^k - \{0\} \times U \to \mathbb{R}^p$ by $F((t_1, \ldots, t_k), v_1, \ldots, v_k) = \sum_{i=1}^{k} t_i v_i$. Show that $F$ is a submersion.

(*iii*) Deduce that if $X \subset \mathbb{R}^p$ is any manifold with $0 \notin X$, then for fixed $k$ almost all $k$-dimensional subspaces $V$ of $\mathbb{R}^p$ meet $X$ transversally. (Apply Thom's Lemma.)

(*iv*) Prove that the result stated in (iii) holds without the restriction $0 \notin X$ iff dim $X + k \geqslant p$.

(2) Prove that Thom's Lemma does not apply to uncountable unions of manifolds $Y$ in $\mathbb{R}^m$. (Consider $G: \mathbb{R} \times \mathbb{R}^2 \to \mathbb{R}^2$ defined by $G(x, a) = a$, with $\bigcup_{a \in \mathbb{R}^2} \{a\}$ the uncountable union.)

(3) Check that in the case where dim $X + $ dim $Y < m$ the proof of Thom's Lemma for a smooth map $G: X \times U \to \mathbb{R}^m \supset Y$ only involves the easy version of Sard's Theorem (namely dim $X < $ dim $Y$ in 8.9). (In particular this was all that was involved in 8.17.)

(4) Let $f: \mathbb{R} \to \mathbb{R}$ be any smooth function and define $\Phi: \mathbb{R} \times \mathbb{R}^2 \to \mathbb{R}$ by $\Phi(t, u_1, u_2) = f(t) + u_1 t + u_2 t^2$. Show that, for all $u = (u_1, u_2) \in \mathbb{R}^2$ outside a null set, the function $\Phi_u: \mathbb{R} \to \mathbb{R}$ has no singularities other than $A_1$ singularities. (Consider $G: \mathbb{R} \times \mathbb{R}^2 \to \mathbb{R}^2$, $G(t, u) = (\Phi'_u(t), \Phi''_u(t))$, and $\{0\} \subset \mathbb{R}^2$. Compare the example following 4.18.)

(5) Let $F: \mathbb{R} \times \mathbb{R}^r \to \mathbb{R}$ be smooth, so that $F$ can be regarded as an $r$-parameter

family of functions $F_x: \mathbb{R} \to \mathbb{R}$, $x \in \mathbb{R}^r$. Define $\Phi: \mathbb{R} \times \mathbb{R}^r \times \mathbb{R}^{r+2} \to \mathbb{R}$ by $\Phi(t, x, u) = F(t, x) + u_1 t + u_2 t^2 + \cdots + u_{r+2} t^{r+2}$. Show that, for all $u \in \mathbb{R}^{r+2}$ outside a null set, the family $\Phi_u: \mathbb{R} \times \mathbb{R}^r \to \mathbb{R}$, $\Phi_u(t, x) = \Phi(t, x, u)$, consists entirely of functions $\Phi_{u,x}: \mathbb{R} \to \mathbb{R}$ having singularities of type $A_{\leqslant r+1}$. (Consider $G: \mathbb{R} \times \mathbb{R}^r \times \mathbb{R}^{r+2} \to \mathbb{R}^{r+2}$, $G(t, x, u) = ((\partial\Phi/\partial t), (\partial^2\Phi/\partial t^2), \ldots, (\partial^{r+2}\Phi/\partial t^{r+2}))$.) Thus *there are families $\tilde{F} = \Phi_u$ arbitrarily close to $F$ which contain only $A_{\leqslant r+1}$ singularities.* (Of course $F$ could equally well have domain $I \times V$ where $I$ is open in $\mathbb{R}$ and $V$ is open in $\mathbb{R}^r$.)

(6) Let $F$ be as in (5). Show that there are families $\tilde{F}$ arbitrarily close to $F$ such that (*i*) $\tilde{F}$ contains only $A_{\leqslant r+1}$ singularities and (*ii*) whenever $\tilde{F}_x$ has type $A_k$ $(k \leqslant r+1)$ at $t$, $\tilde{F}$ is a versal unfolding of $\tilde{F}_x$ at $t$. (Compare 6.21.)

(7) Using the ideas of 8.18 show that, given smooth manifolds $X$ in $\mathbb{R}^n$ and $Y$ in $\mathbb{R}^m$ and a smooth map $f: X \to \mathbb{R}^m$ there exist arbitrarily close approximations to $f$ which are transverse to $Y$.

## Algebraic surfaces and more on apparent contours

'A long shot, Watson; a very long shot!'
(*Silver Blaze*)

Our final application of Thom's Lemma in this chapter deals with the projections of surfaces to a plane. Let $Q_d$ denote the real vector space of all polynomials $F$ of degree $\leqslant d$ in three variables $t, x_1, x_2$. Sard's Theorem would lead us to hope that most of the algebraic surfaces $F = 0$ are smooth, and the reasoning of 7.13 would then encourage us to believe that the projection of $F^{-1}(0)$ to the $(x_1, x_2)$-plane would have the local forms of fig. 7.3 for 'most' $F$. We shall actually prove that this is so, at any rate when $d \geqslant 3$. (The case $d = 2$ is almost done in 7.13, and the details are left to the reader.)

**8.21     Theorem**

*For almost all $F \in Q_d$ (in the sense of Lebesgue measure) the algebraic surface $F = 0$ is smooth, and the projection $(t, x) \mapsto x$ is locally one of the forms discussed in 7.13. (In particular the apparent contour is locally either smooth or an ordinary cusp.)*

**Proof** The idea of the proof is quite simple. Define $\Phi_i: \mathbb{R}^3 \times Q_d \to \mathbb{R}^4$, $i = 1, 2, 3$ by

$$\Phi_1((t, x), F) = \left( F, \frac{\partial F}{\partial t}, \frac{\partial F}{\partial x_1}, \frac{\partial F}{\partial x_2} \right)(t, x),$$

$$\Phi_2((t, x), F) = \left( F, \frac{\partial F}{\partial t}, \frac{\partial^2 F}{\partial t^2}, \frac{\partial^3 F}{\partial t^3} \right)(t, x),$$

$$\Phi_3((t, x), F) = \left( F, \frac{\partial F}{\partial t}, \frac{\partial^2 F}{\partial t^2}, \frac{\partial F}{\partial x_1} \frac{\partial^2 F}{\partial t \partial x_2} - \frac{\partial F}{\partial x_2} \frac{\partial^2 F}{\partial t \partial x_1} \right)(t, x).$$

If we can show that 0 is a regular value of each $\Phi_i$ then by Thom's Lemma it will follow that, for almost all $F$, $\Phi_i(-, F): \mathbb{R}^3 \to \mathbb{R}^4$ will have 0 as regular value, that is $\Phi_i((t, x), F) = 0$ will have no solution. The discussion of 7.13 then proves the required result.

The first two maps $\Phi_1, \Phi_2$ present little difficulty. Suppose for $\Phi_2$ that we consider tangent vectors from the space $Q_d$ obtained from the paths $s \mapsto F_s$ where $F_s = F + s, F + st, F + st^2, F + st^3, s \in (-1, 1)$. (This is where we need $d \geqslant 3$.) The images of these tangent vectors under the derivative $D\Phi_2$ are the vectors

$$\lim_{s \to 0} \frac{\Phi_2((t, x), F_s) - \Phi_2((t, x), F)}{s}$$

(compare 4.23(3)). These are respectively $(1, 0, 0, 0)$, $(t, 1, 0, 0)$, $(t^2, 2t, 2, 0)$, $(t^3, 3t^2, 6t, 6)$, so clearly $D\Phi_2$ is surjective at every point, i.e. $\Phi_2$ is a submersion.

The same reasoning (using $F_s = F + s, F + st, F + sx_1, F + sx_2$) shows that $\Phi_1$ is also a submersion. The problem with $\Phi_3$ however comes from the product terms. By considering the paths $F_s = F + s, F + st, F + st^2, F + sx_1, F + sx_2, F + stx_1, F + stx_2$ one obtains the tangent vectors $(1, 0, 0, 0)$, $(t, 1, 0, 0)$, $(t^2, 2t, 2, 0)$ and $(x_1, 0, 0, \partial^2 F/\partial t \partial x_2)$, $(x_2, 0, 0, -\partial^2 F/\partial t \partial x_1)$, $(tx_1, x_1, 0, t\partial^2 F/\partial t \partial x_2 - \partial F/\partial x_2)$, $(tx_2, x_2, 0, \partial F/\partial x_1 - t\partial^2 F/\partial t \partial x_1)$. Clearly we now have a submersion unless $\partial F/\partial x_1 = \partial F/\partial x_2 = \partial^2 F/\partial t \partial x_1 = \partial^2 F/\partial t \partial x_2 = 0$. We can circumvent this embarrassing problem by the following piece of low cunning. Now 0 is a regular value of $\Phi_1$ so $X = \Phi_1^{-1}(0)$ is a smooth closed manifold in $\mathbb{R}^3 \times Q_d$ of codimension 4. We claim that $\Phi_3$ restricted to the open set $\mathbb{R}^3 \times Q_d - X$ has 0 as a regular value. (Proof: if $D\Phi_3(t, x, F)$ is not surjective and $\Phi_3(t, x, F) = 0$ then $\partial F/\partial x_1 = \partial F/\partial x_2 = F = \partial F/\partial t = 0$, so $(t, x, F) \in X$.) Let $\pi$ be the natural projection of $\mathbb{R}^3 \times Q_d$ to $Q_d$.

A small adaptation of the argument of 8.17 shows that there is a set $\Omega$ of measure zero in $Q_d$ such that, for all $F \in Q_d - (\pi(X) \cup \Omega)$, the map $\Phi_3(-, F)$, defined on all of $\mathbb{R}^3$, has 0 as a regular value. However $\dim X < \dim Q_d$ so $\pi(X)$ consists of the critical values of $\pi | X$ and so has measure zero in $Q_d$ by Sard's theorem. Hence $\Phi_3(-, F): \mathbb{R}^3 \to \mathbb{R}^4$ has 0 as a regular value for almost all $F \in Q_d$.

It now follows that, for almost all $F \in Q_d$, 0 is a regular value of all three maps $\Phi_i(-, F)$, and that gives the required result.     □

It is well worth noting that the method of 8.21 proves a general result about smooth functions $F: \mathbb{R}^3 \to \mathbb{R}$ too. Let $F$ be such, and write $P$ for an arbitrary element of $Q_d$, where $d \geqslant 3$. Then new maps $\Phi_i$ are given by setting $\Phi_i((t, x), P)$ equal to the old expression with $F + P$ substituted for $F$. The conclusion is now that, *for almost all $P \in Q_d$, $F + P$ has 0 as a regular value and the projection $(t, x) \mapsto x$ is locally one of the forms of* 7.13, *so that the apparent contour of $(F + P)^{-1}(0)$ in the $t$-direction is locally smooth or a cusp.* In particular $P$ can be chosen arbitrarily close to the zero polynomial: the above can be achieved by an arbitrarily small cubic deformation of $F$.

## Open and dense properties

We have been trying to convince you that transversality is a typical property, that is that it holds nearly always. We would like to be more precise about what 'nearly always' really means. To do this we make the following definitions. Below $X$ and $Y$ are smooth manifolds, as usual.

### 8.22    Definitions

(1) A property $P$ of smooth mappings $f: X \to Y$ is said to be *dense* if for any such $f$ the following holds. There exists a family of smooth mappings $F: X \times U \to Y$, with $U$ some neighbourhood of $0 \in \mathbb{R}^N$ (some $N \geqslant 1$), with $F_0 = f$ (i.e. $F(x, 0) = f(x)$) and with $F_a = F(-, a)$ having property $P$ for $a$ arbitrarily close to 0.

(2) A property $P$ of smooth mappings $f: X \to Y$ is said to be *open* if for any such $f$ satisfying property $P$ the following holds. Given any neighbourhood $U$ of $0 \in \mathbb{R}^N$ ($N \geqslant 1$) and any family $F: X \times U \to Y$, with $F_0 = f$, $F_a$ has property $P$ for all $a$ sufficiently close to $0 \in U$.

Intuitively these definitions are interpreted as follows. A property $P$ is dense if any map $f$ can be arbitrarily closely approximated by a map possessing property $P$. A property $P$ is open if given any map $f$ with property $P$ then any other map $g$ sufficiently close to $f$ in a finite dimensional family will also have property $P$. We think of a property which is open and dense as one which holds nearly always. (Although we do take this as our definition of 'nearly always' there are subtle objections – see 8.24(3) below.)

Clearly Thom's Lemma is designed to prove that certain transversality properties are dense. (Indeed the conclusion one can draw from Thom's Lemma is considerably stronger: the good values of $a$ not only contain 0 as an accumulation point, their complement has measure zero in $U$.) Although there are fairly general results which assert that transversality is a dense property by and large each situation requires a different application of Thom's Lemma. What we *shall* prove is:

### 8.23     Proposition

*Let $X$ be compact and $Y \subset \mathbb{R}^m$ be a smooth manifold which is a closed subset of $\mathbb{R}^m$, and consider the set of smooth mappings $f: X \rightarrow \mathbb{R}^m$. The property 'f is transverse to Y' is open.*

***Proof*** Let $F: X \times U \rightarrow \mathbb{R}^m$ be any smooth family with $F_0 = f$ and $f \pitchfork Y$. Using the compactness of $X$ we need only show that each point $(x', 0) \in X \times U$ has an open neighbourhood $V$ in $X \times U$ with $F_a \pitchfork Y$ at $x$ for all $(x, a) \in V$. (The union of all such sets $V$ covers the compact set $X \times \{0\}$ and so there is a finite cover $V_1, \ldots, V_n$ of this set. If $\pi: X \times U \rightarrow U$ is the natural projection the sets $\pi(V_i)$ are also open and $\bigcap_{i=1}^{n} \pi(V_i)$ gives a neighbourhood $U'$ of $0 \in U$ with $F_a \pitchfork Y$ for $a \in U'$.)

We now construct the set $V$. Choose a neighbourhood $W$ of $f(x')$ so that if $f(x') \notin Y$ then $W \cap Y = \varnothing$ (this uses $Y$ closed in $\mathbb{R}^m$), and if $f(x') \in Y$ there is a smooth map $g: W \rightarrow \mathbb{R}^k$, with $0$ a regular value of $g$ and $g^{-1}(0) \cap W = Y \cap W$. We now set $V_1 = F^{-1}(W)$. Clearly if $f(x') \notin Y$ this will do for our open set $V$. If $f(x') \in Y$ we know that the restriction of $g \circ F$ to $X \times \{0\} \cap V_1$ is a submersion at $(x', 0)$, since $g \circ F_0 = g \circ f$ and $f$ is transverse to $Y$. Hence $T(g \circ F_0)(x')$ is a surjective linear map. So for $(x, a)$ in some neighbourhood $V$ of $(x', 0)$, with $V \subset V_1$, the linear map $T(g \circ F_a)(x)$ will still be surjective (for the appropriate matrix will still have rank $k$, being close to a matrix of rank $k$). This means that $F_a$ is transverse to $Y$ at $x$ for any $(x, a) \in V$, whence the result.     $\square$

### 8.24     Exercise

(1) Show that the condition $Y$ is closed in 8.23 cannot be dropped by considering $X = S^1$ the unit circle, $Y = \{(x_1, x_2, 0) \in \mathbb{R}^3: x_1^2 + x_2^2 < 1\}$ and $f: X \rightarrow \mathbb{R}^3$ the map $f(x_1, x_2) = (x_1, x_2, 0)$.

(2) Show that the condition $X$ is compact in 8.23 cannot be dropped, or even replaced by $X$ is closed. (Reverse the roles of $X$ and $Y$ in (1), noting that $Y$ is diffeomorphic to the whole of $\mathbb{R}^2$ (a closed set).)

(3) Show that open dense sets in $\mathbb{R}^n$ need not be very large. More precisely given any $\varepsilon > 0$ produce an open dense set $X$ in $\mathbb{R}^n$ of measure $< \varepsilon$. (Only those with some knowledge of Lebesgue measure need try this one.) This shows that one must be careful about making extravagant claims for open dense properties.

(4) Let $X$ and $Y$ be smooth compact manifolds. Show that the property '$f: X \rightarrow Y$ is a diffeomorphism' is open but never dense.

**8.25   Project**

(1) Another interesting use of Sard's Theorem is to show that a smooth
$n$-manifold $M$, initially in say $\mathbb{R}^N$ for some large $N$, can be realized in
$\mathbb{R}^{2n+1}$. Given a unit vector $a \in \mathbb{R}^N$ the orthogonal projection $\pi_a: \mathbb{R}^N \to \mathbb{R}^N$ to
the hyperplane perpendicular to $a$ is given by $\pi_a(x) = x - (x \cdot a)a$.

(i) Show that the tangent bundle of $M$, namely

$$TM = \{(x, v) \in M \times \mathbb{R}^N : v_x \in M_x\}$$

is a smooth $2n$-manifold in $\mathbb{R}^{2N}$.

(ii) Let $\phi_1: TM \to \mathbb{R}^N$, $\phi_2: M \times M \times \mathbb{R} \to \mathbb{R}^N$ be defined by $\phi_1(x, v) = v$,
$\phi_2(x_1, x_2, t) = t(x_1 - x_2)$. Show that if $a \in \mathbb{R}^N$ is not in the image of $\phi_1$
(resp. $\phi_2$) then $\pi_a$ gives a smooth map $\pi_a | M: M \to \mathbb{R}^N$ which has maximal
rank at every point of $M$ (resp. is injective).

(iii) Assuming $N > 2n+1$, use Sard's Theorem to deduce that $a$ can be
chosen so that $\pi_a | M$ is of maximal rank everywhere and injective.

When $M$ is compact, this actually shows that $\pi_a(M)$ is also a smooth
$n$-manifold, in the hyperplane perpendicular to $a$, so essentially in $\mathbb{R}^{N-1}$:
$M$ has been projected diffeomorphically into $\mathbb{R}^{N-1}$. Repeating the argument gets it down to $\mathbb{R}^{2n+1}$. Compare Guillemin and Pollack (1974),
chapter 1.

(2) One can start from a more general idea of smooth manifold (or differentiable manifold) as a topological space (Hausdorff with countable basis)
satisfying suitable axioms. It is then possible to prove that any smooth
$n$-manifold $M$ can be realized (embedded) in a Euclidean space $\mathbb{R}^N$ for some
$N$. As in (1) this can then be brought down to $\mathbb{R}^{2n+1}$. You can read about
this in Bröcker and Jänich (1982) or, in the compact case, in chapter 1 of
Hirsch (1976).

# 9

## Generic properties of curves

'You know my methods. Apply them!'
(*The Hound of the Baskervilles*)

In this chapter we shall associate to plane (and space) curves other curves in some Euclidean space from which we can read off certain vital aspects of the geometry of the original curve. For example for the limaçon we have the following picture (fig. 9.1).

From the associated curve (on the right) we can deduce that the limaçon has four vertices and two inflexions, and the inflexions are all ordinary. This follows from the way the curve meets the $a_2$ and $a_3$ axes.

Using these associated curves and Thom's Transversality Lemma (chapter 8) we shall justify a number of assertions made in previous chapters (in particular chapter 7). For example we shall prove that 'almost all' compact plane curves have only ordinary inflexions and

**Fig. 9.1.**

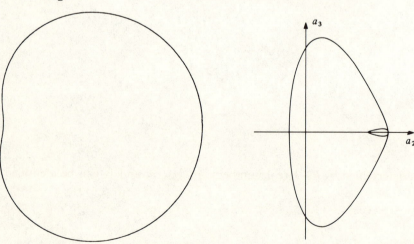

vertices. It then follows from our results in chapter 7 that we can describe the local appearance of the dual, evolute and Gauss map of almost any compact plane curve.

Our first task is to extract from a plane curve the infinitesimal information which is needed to determine vertices, inflexions and the like. This we do using the Monge–Taylor mapping – it is this mapping which gives rise to the auxiliary curves mentioned above. (In a later section we shall describe the analogous construction for space curves.)

In this chapter we shall assume more familiarity with basic topological notions than previously. Also not all of the results will be proved in full detail. In particular the final section concerning genericity and space curves is rather brief and the reader will need a good understanding of the previous material in order to follow this discussion.

## The Monge–Taylor map

From 9.5 onwards we shall usually consider only compact curves, that is images of a circle rather than an open interval. Let $S^1 \subset \mathbb{R}^2$ be the unit circle and let $\theta: \mathbb{R} \to S^1$ be $\theta(s) = (\cos s, \sin s)$. Then $\gamma: S^1 \to \mathbb{R}^n$ is smooth if and only if $\gamma \circ \theta: \mathbb{R} \to \mathbb{R}^n$ is smooth (exercise 8.10(1)); we say that $\gamma$ is a *regular* (parametrized) *curve* provided $\gamma \circ \theta$ is a regular curve. Increasing $s$, i.e. anticlockwise orientation of $S^1$, gives a well-defined tangent and normal at each point $\gamma(t)$, $t \in S^1$, despite the fact that there are infinitely many corresponding values of $s$. When $I$ is a connected open subset of $S^1$ then $\gamma: I \to \mathbb{R}^n$ is a *regular curve* provided, for each component $J$ of $\theta^{-1}(I)$, $\gamma \circ \theta: J \to \mathbb{R}^n$ is a regular curve. These curves $\gamma$, when $I \neq S^1$, are no different from those considered so far in this book; from now on we shall regard the domain $I$ of a regular curve as any connected open subset of $S^1$.

Let $\gamma: I \to \mathbb{R}^2$ be a (regular) plane curve. We shall associate to each point $t \in I$ a polynomial of degree $k$ which carries all of the infinitesimal information concerning the curve $\gamma$ at $t$ up to order $k$. Let $V_k(k \geq 2)$ be the space of polynomials in a single variable $\xi$ of degree $\leq k$ and $\geq 2$. So each element of $V_k$

$$a_2\xi^2 + a_3\xi^3 + \cdots + a_k\xi^k$$

is determined by $k - 1$ real numbers, $a_2, \ldots, a_k$ and $V_k$ can be identified with $\mathbb{R}^{k-1}$ via the coordinates $(a_2, \ldots, a_k)$. Given $t \in I$ there is a unit tangent vector $T(t)$, and unit normal vector $N(t)$ to the curve at the point with parameter $t$. Using these two directions as axes we can write $\gamma(I)$ locally at $\gamma(t)$ as the graph of a function $\eta = f_t(\xi)$, i.e. in *Monge normal form*, where $\xi$ is along the $T$ direction, $\eta$ along the $N$ direction (fig. 9.2). Clearly $f_t$ is uniquely determined by $t$. Note that it is also unaffected by rigid motions of $\gamma(I)$ in the plane, and, if $\gamma \circ h$ is a reparametrization of $\gamma$ (as in 2.16), then $f_t = f_s$ where $h(s) = t$.

## 9.1     Definition

The *Monge–Taylor map* (of order $k$) $\mu_\gamma : I \to V_k$ is the map which associates to each $t \in I$ the $k$-jet of the function $f_t$ at 0. Thus $\mu_\gamma(t) = j^k f_t(0)$, which is the Taylor series of $f_t$ at 0 truncated to degree $k$.

We often write just $\mu$ instead of $\mu_\gamma$, and identify $V_k$ with $\mathbb{R}^{k-1}$ so that $\mu_\gamma(t)$ has the form $(a_2(t), \ldots, a_k(t))$ where $a_i(t) = f_t^{(i)}(0)/i!$. Note that since the $\xi$-axis is tangent to $C$ at $\gamma(t)$ we clearly have $f_t(0) = f_t'(0) = 0$ (where $'$ denotes $d/d\xi$) so the image of $\mu_\gamma$ is indeed a subset of $V_k$.

## 9.2     Example

Consider the unit circle $\gamma(t) = (\cos t, \sin t)$ and the point $(0, -1)$ corresponding to $t = 3\pi/2$ (fig. 9.3). In $\xi, \eta$ coordinates at $(0, -1)$ the circle has equation $\xi^2 + (\eta - 1)^2 = 1$, i.e. $\xi^2 + \eta^2 - 2\eta = 0$, so $\eta = 1 \pm (1 - \xi^2)^{\frac{1}{2}}$. We are concerned with that part of the circle given by $\eta = 1 - (1 - \xi^2)^{\frac{1}{2}} = \frac{1}{2}\xi^2 + \frac{1}{8}\xi^4 + \cdots$ for small $\xi$, and $\mu(3\pi/2) = j^k(1 - (1 - \xi^2)^{\frac{1}{2}})(0)$. Because of symmetry the image of the whole of the circle consists of this point, which for $k = 3$ is $(\frac{1}{2}, 0)$.

It is clear that the Monge–Taylor map $\mu_\gamma$ does carry all of the infinitesimal information concerning $\gamma$ at each point $\gamma(t)$ up to $k$th order. At each point $\gamma(t)$ the numbers $a_2(t), \ldots, a_k(t)$ can be expressed in terms of the

**Fig. 9.2.** Monge-Taylor map

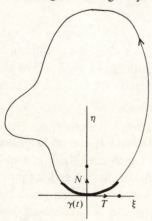
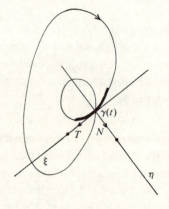

**Fig. 9.3.** Monge–Taylor map for a circle

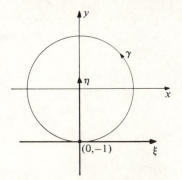

curvature of $\gamma$ at $t$ and the first $k-2$ derivatives of the curvature with respect to the tangential coordinate $\xi$. This follows directly from 2.28(7); it also follows that we can make $\gamma$ unit speed and use derivatives of curvature with respect to arclength instead. The first few formulas, in 9.3(*i*) below, follow from 2.28(8) ($Y^{(i)}(0)$ there is our $f_t^{(i)}(0)=i!a_i(t)$). Part (*ii*) of 9.3 follows from (*i*) or rather more directly from 2.15(1).

**9.3    Proposition**

   (*i*)  *With* $' = d/d\xi$ *we have* $\kappa(t)=2a_2(t)$, $\kappa'(t)=6a_3(t)$, $\kappa''(t)=24(a_4(t)-(a_2(t))^3)$. (*ii*) $\gamma$ *has an inflexion at t iff* $a_2=0$, *a higher inflexion iff* $a_2=a_3=0$, *a vertex\* iff* $a_2\neq0$, $a_3=0$ *and a higher vertex iff* $a_2\neq0$, $a_3=a_4-a_2^3=0$.

$\square$

Let us take $k=3$ and consider the Monge–Taylor map $\mu\colon I\to V_3$. The target $V_3$ can be thought of as a plane with coordinates $(a_2, a_3)$ and the image of $\mu$ as a curve in this plane (the curve might collapse as in the case of a circle). If $\mu(I)$ passes through the origin, by 9.3(*ii*) $\gamma$ has a higher inflexion; if $\mu(I)$ crosses the $a_2$-axis, not at $(0, 0)$, then $\gamma$ has a vertex and so on. In fig. 9.1 at the beginning of this chapter we have drawn a limaçon and the image of its Monge–Taylor map (when $k=2$). Since the Monge–Taylor map cuts the $a_3$-axis twice (not through the origin) and the $a_2$-axis four times (twice at the same place) we can deduce that the limaçon has two ordinary inflexions and four vertices. Fig. 9.4 shows the Monge–Taylor curve associated to an ellipse. This time the associated curve is traversed *twice* as we proceed once around the ellipse. (This is because of the reflective symmetry of the ellipse of course: at points that are reflexions in the centre of the ellipse, $a_2$ and $a_3$ (and all higher $a_i$) coincide.)

---

\* In some books $a_2=a_3=0$ would count as a vertex. See the footnote on p. 18.

**Fig. 9.4.** Image of Monge–Taylor map for an ellipse ($k = 3$)

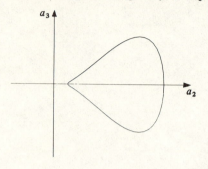

One last, rather interesting point. It is natural to ask to what extent the Monge-Taylor maps determine the curve $\gamma$. Recalling the result (mentioned in 2.37) that the curve is determined essentially by the curvature function, and using the fact that $\kappa(t) = 2a_2(t)$, one is tempted to deduce that the curve is determined up to a rigid motion by the map $\mu: I \to V_2$. However the curve is only determined up to rigid motion if the curvature is given *as a function of arc-length s*, so this reasoning is fallacious. You might however like to prove that $\gamma$ *is* determined up to rigid motion by $\mu: I \to V_3$. (Hint: $\kappa(t) = 2a_2(t)$, and if $s = l(t)$ is the arc-length function, $d/ds \, (\kappa \circ l^{-1})(s) = 6a_3(l^{-1}(s))$. This determines $l'(t)$ and hence (essentially) $\kappa \circ l^{-1}$ as required.)

## 9.4    Exercises

(1) Let $\gamma: I \to \mathbb{R}^2$ be a (regular) curve, not necessarily unit speed, with $\gamma(t_0) = 0$. Then the coordinates $\xi$, $\eta$ of $\gamma(t)$ relative to axes $T(t_0)$, $N(t_0)$ are functions of $t$: $\xi(t) = \gamma(t) \cdot T(t_0)$, $\eta(t) = \gamma(t) \cdot N(t_0)$, $\eta(t) = f(\xi(t))$ where $f = f_{t_0}$ in the notation of 9.1. Differentiating these equations and using $a_i = f^{(i)}(t_0)/i!$ show that

$$2a_2 = (\gamma' \cdot T)^{-2}(\gamma'' \cdot N),$$
$$6a_3 = (\gamma' \cdot T)^{-4}((\gamma' \cdot T)(\gamma''' \cdot N) - 3(\gamma'' \cdot T)(\gamma'' \cdot N)),$$
$$24a_4 = (\gamma' \cdot T)^{-6}((\gamma' \cdot T)^2(\gamma^{(4)} \cdot N) - 6(\gamma' \cdot T)(\gamma''' \cdot N) + 15(\gamma'' \cdot T)^2(\gamma'' \cdot N)$$
$$- 4(\gamma' \cdot T)(\gamma'' \cdot N)(\gamma''' \cdot T))$$

where everything is evaluated at $t_0$.

(2) Let $\gamma(t) = (t, at^2)$ where $a$ is constant. For each $t_0 \in \mathbb{R}$ apply (1) to $\gamma(t) - \gamma(t_0)$ and show that for all $t \in \mathbb{R}$,

$$a_2(t) = \frac{a}{(1 + 4a^2t^2)^{\frac{3}{2}}}, \; a_3(t) = \frac{-4a^3t}{(1 + 4a^2t^2)^3}, \; a_4(t) = \frac{20a^5t^2}{(1 + 4a^2t^2)^{\frac{9}{2}}}.$$

(3) Let $\mu: I \to V_3$ be the Monge–Taylor map of $\gamma$ and let $t \in I$. Use 9.3($i$) to show that

$$\mu'(t) = l'(t)(3a_3(t), 4(a_4(t) - (a_2(t))^3))$$

where $l$ is the arc-length function for $\gamma$ (so $l'(t)=1$ if $\gamma$ is unit speed). Use this to show $(i)$ if $\gamma$ has an ordinary inflexion at $t$ then $\mu$ is transverse to the inflexion set $a_2=0$ at $t$, $(ii)$ if $\gamma$ has an ordinary vertex at $t$ then $\mu$ is transverse to the vertex set $a_3=0$ at $t$, $(iii)$ $\mu$ gives a regular curve in $V_3=\mathbb{R}^2$ iff all vertices of $\gamma$ are ordinary.

(4) Let $\mu: I \to V_3$ be as before and suppose $\mu(I)$ consists of a single point. Show that $\gamma(I)$ is a circular arc or a straight segment in $\mathbb{R}^2$. You can assume $\gamma$ is unit speed.

(5) Suppose $\gamma$ is unit speed and $\mu: I \to V_3$ satisfies $a_3(t)+ba_2(t)+c=0$ for constants $b$ and $c$. Find the 'intrinsic equation' of $\gamma$, i.e. express the tangent angle $\psi$ as a function of arc-length $t$ (compare 2.24). (Use the fact that, in 9.3$(i)$, we can interpret $'$ as d/d$t$ since $\gamma$ has unit speed.)

(6) Let $(x_0, y_0)$ be a point of a plane curve $C$ given by $g(x, y)=0$, where 0 is a regular value of $g$. We suppose $C$ oriented near $(x_0, y_0)$, so that a definite direction is chosen for the tangent vector $T=(T_1, T_2)$ at this point. Show that $T=\pm(-g_y, g_x)/(g_x^2+g_y^2)^{\frac{1}{2}}$. Here $g_x$ stands for $\partial g/\partial x\,(x_0, y_0)$; similarly $g_{xy}$ will be $\partial^2 g/\partial x\partial y(x_0, y_0)$, and so on.

Supposing the upper sign gives the correct $T$, show that
$$2a_2(g_x^2+g_y^2)^{\frac{1}{2}}=g_{xx}T_1^2+2g_{xy}T_1T_2+g_{yy}T_2^2.$$
(One way of doing this which produces formulae for the higher $a_i$ is this. Show $(x, y)=(x_0+\xi T_1-\eta T_2, y_0+\xi T_2+\eta T_1)$ are the ordinary coordinates of the point whose coordinates relative to tangent and normal axes at $(x_0, y_0)$ are $(\xi, \eta)$. Write formally $\eta=a_2\xi^2+a_3\xi^3+\cdots$, supposing $g(x, y)=0$ as well as $g(x_0, y_0)=0$. Expand $g(x, y)$ using Taylor's theorem in two variables and compare coefficients of $\xi^2$ on the two sides.)

(7) Use the above formula to find $a_2$ at $(x_0, y_0)$ for the ellipse $x^2/a^2+y^2/b^2=1$ and the oval $y^2=a^2(x-x^3)(0\leqslant x\leqslant 1)$, both oriented anticlockwise. Compare 2.26(1), (3).

(8) Find the formula for $a_3$ at $(x_0, y_0)$ in terms of derivatives of $g$ at $(x_0, y_0)$ and components of $T$. If your formula involves $a_2$ as well, that does not matter. (These formulae for $a_i$ are very convenient for computing values numerically. The formulae for $a_2$, $a_3$, $a_4$ were used by the authors in producing computer pictures of curves associated with caustics by reflexion, in particular the 'focoid' of 7.21(15).)

We shall now attempt to outline our proof of the fact that almost any compact plane curve $\gamma: S^1 \to \mathbb{R}^2$ has only ordinary inflexions and vertices. Above we remarked that a curve $\gamma: I \to \mathbb{R}^2$ has a higher inflexion at $t$ iff $\mu(t)$ is the zero polynomial (i.e. the origin) in $V_3$. But supposing $\mu(I)$ does pass through the origin we would expect that by slightly deforming $\gamma$, to $\gamma_1$ say, the corresponding Monge–Taylor map $\mu_1: I \to V_3$ could be made to miss the origin in the $V_3$ plane.

Likewise with $k=4$ a plane curve $\gamma: I \to \mathbb{R}^2$ gives rise to a space curve $\mu: I \to V_4$, with coordinates $(a_2, a_3, a_4)$ (the image of $\mu$ might again collapse of course). The set $a_3 = a_4 - a_2^3 = 0$ in $V_4$ of higher vertices is itself a curve (see fig. 9.5) and 'in general' we would expect $\mu(I)$ to miss this curve. Again if $\mu(I)$ did happen to meet the curve of higher vertices one would hope that a slight deformation of $\gamma$ to $\gamma_1$ would result in a map $\mu_1 : I \to V_4$ which did miss this curve.

It should be clear from our work in chapter 8 that what we require is that $\mu$ should be transverse to the relevant subsets of $V_3$ and $V_4$. The key to the proof then is going to be a transversality theorem. There is however a major flaw in the plausible argument given above. It is far from clear how changes in $\gamma$ are going to affect the map $\mu$, so that for instance it is not clear (and in fact not true) that one can deform $\mu$ to avoid the sets described above and *afterwards* obtain a corresponding deformation of $\gamma$. There are indeed 'hidden' constraints on the map $\mu$; for instance when $k = 3$ the curve, $\mu(I)$, cannot touch the $a_3$-axis except at the origin (can the reader see why?), while the 'four vertex theorem' (DoCarmo (1976)) implies that for a compact (plane) curve at least four points of $I$ must map to points of the $a_2$-axis.

To overcome this objection we have to show that there are sufficiently many deformations of our original curve $\gamma$ to give all of the deformations we require of the Monge–Taylor map $\mu$. This we do as follows.

Let $P_k$ denote the set of maps $\psi : \mathbb{R}^2 \to \mathbb{R}^2$ of the form $\psi(x, y) = (\psi_1(x, y), \psi_2(x, y))$ where each $\psi_i$ is a polynomial in $x$ and $y$ of degree $\leqslant k$. So an element $\psi \in P_k$ is determined by the coefficients of the various monomials $x^i y^j$ occuring in $\psi_1$ and $\psi_2$. There are altogether $1 + 2 + \cdots + (k+1) = \frac{1}{2}(k+1)(k+2)$ monomials of degree $\leqslant k$, so $P_k$ can be thought of as a Euclidean space $\mathbb{R}^N$ with $N = (k+1)(k+2)$. It is this space which will provide the required deformations of the curve.

To simplify matters *we now assume that the curve $\gamma(I)$ is compact*, i.e. $I = S^1$. The identity map $\mathrm{id} : \mathbb{R}^2 \to \mathbb{R}^2$, $\mathrm{id}(x, y) = (x, y)$, is of course an element

**Fig. 9.5.** The higher vertex set in $V_4$

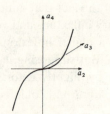

of $P_k$ (provided $k \geqslant 1$), and using the compactness of $\gamma(S^1)$ it easily follows that there is a neighbourhood $U$ of id in $P_k$ with the property that if $\psi \in U$ then $\psi \circ \gamma \colon S^1 \to \mathbb{R}^2$ is a regular curve. The key result of this chapter is the following.

**9.5    Theorem**

  *Let $Q$ be a manifold in $V_k = \mathbb{R}^{k-1}$. For some open set $U_1 \subset U$ containing the identity map the map $\mu \colon S^1 \times U_1 \to V_k$ defined by $\mu(t, \psi) = \mu_{\psi \circ \gamma}(t)$ is transverse to $Q$.* (In fact we shall prove that $\mu$ is a submersion so $Q$ does not enter the argument at all.)

***Proof*** Using the compactness of $S^1$ and the fact that being a submersion is an open condition it is enough to consider a point $(t, \text{id}) \in S^1 \times U$. (Compare the proof of 8.23.) By choosing our coordinates we can further assume that $\gamma(t) = 0 \in \mathbb{R}^2$ and that close to $0$ the curve is given by the equation $y = f(x)$, where $f(0) = f'(0) = 0$, so $\mu(t, \text{id}) = j^k f(0)$. ('Choosing coordinates' means that we first do a rigid motion $\theta$ of $\mathbb{R}^2$ to bring $\gamma(S^1)$ into this standard position, with $\gamma(t)$ moved to $(0, 0)$ and the tangent line moved to the $x$-axis. This does not affect the result because referring our transformations $\psi$ to the new coordinates leaves them of the same degree, and the Monge–Taylor expansion of $\gamma(S^1)$ at $\gamma(t)$ is the same as that of $\theta(\gamma(S^1))$ at $\theta(\gamma(t))$.)

  Let $F(x)$ be a polynomial in $x$ of degree $\geqslant 2$ and $\leqslant k$ and let $\psi^s(x, y) = \psi(s, x, y) = (x, y + sF(x))$ where $s \in \mathbb{R}$. Then, for all small $s$, $\psi^s \in U$ and we consider the path $s \mapsto (t, \psi(s, x, y))$ in $S^1 \times U$. Now since $\psi^s(s, u, f(u)) = (u, f(u) + sF(u))$ and the degree of $F$ is at least $2$ the Monge–Taylor expansion of $\psi^s(\gamma(S^1))$ at $\psi^s(\gamma(t)) = 0$ is $j^k(f + sF)(0)$. The corresponding tangent vector in $V_k$ at $\mu(t, \text{id})$ is

$$\lim_{s \to 0} \frac{j^k(f + sF)(0) - j^k f(0)}{s} = \lim_{s \to 0} \frac{j^k f(0) + s j^k F(0) - j^k f(0)}{s} = j^k F(0) = F(\xi).$$

(This method of obtaining tangent vectors was developed in 4.23(3).)

  Since $F$ was *any* polynomial with $2 \leqslant \text{degree } F \leqslant k$ this clearly proves that $\mu$ is a submersion at $(t, \text{id})$ and the result follows.    □

It should now be clear how we want to apply the theorem above: for $Q$ we select the submanifold of $V_3$ (resp. $V_4$) consisting of higher inflexions (respectively higher vertices). We now apply Thom's Lemma 8.17 to the family $\mu \colon S^1 \times U_1 \to V_3$ (resp. $V_4$). Thom's Lemma assures us that for a dense set of $\psi \in U_1$ the map $\mu_{\psi \circ \gamma} \colon S^1 \to V_3$ (resp. $V_4$) will be transverse to $Q$. But codim $Q > 1$ so transversality implies that $\mu_{\psi \circ \gamma}(S^1)$ misses $Q$. Hence the curves $\{\psi \circ \gamma\}$ for $\psi$ in this dense subset of $U_1$ will not have higher inflexions (resp. higher vertices). In particular then we can find a sequence of polynomial mappings $\psi_n$ converging to the identity id with $\{\psi_n \circ \gamma\}$ having these properties for all $n$.

Before stating the precise result we want we need to borrow some nomenclature from chapter 8 (see 8.22). For clarity we repeat the definitions in the current notation.

### 9.6    Definitions

(i) A property $P$ is said to be *dense* or to *hold for a dense set of* (*regular*) *plane curves* $\gamma: I \to \mathbb{R}^2$ if the following holds. For any such $\gamma$ there should be an open neighbourhood $U$ of 0 in some Euclidean space $\mathbb{R}^N$, and a family of regular plane curves $\tilde{\gamma}: I \times U \to \mathbb{R}^2$ such that

(a) $\tilde{\gamma}(t, 0) = \gamma(t)$,

(b) for some sequence $\{u_n\}$ in $U$ with $\lim_{n \to \infty} u_n = 0$ property $P$ holds for the sequence of curves $\gamma_n$ defined by $\gamma_n(t) = \tilde{\gamma}(t, u_n)$.

(ii) A property $P$ is said to be *open* or to *hold for an open set of* (*regular*) *plane curves* if given such a curve $\gamma: I \to \mathbb{R}^2$ with property $P$ and any family $\tilde{\gamma}: I \times U \to \mathbb{R}^2$ of (regular) curves $\tilde{\gamma}_u$ (where $\tilde{\gamma}_u(t) = \tilde{\gamma}(t, u)$) the property $P$ holds for all curves $\tilde{\gamma}_u$ with $u$ in some neighbourhood $U_1$ of 0.

(iii) A property $P$ is said to be *generic* or to hold for a *generic set of curves* if it is both open and dense.

The main result is the following:

### 9.7    Corollary

*An open dense set of regular plane curves $\gamma: S^1 \to \mathbb{R}^2$ have only finitely many ordinary inflexions and ordinary vertices, and no higher inflexions or higher vertices. Thus these properties are generic.*

It follows at once that for an open dense set of (regular) curves $\gamma: S^1 \to \mathbb{R}^2$ the evolute (resp. dual) is locally a regular curve except for the presence of finitely many ordinary cusps corresponding to the vertices (resp. inflexions) of $\gamma$. Compare 7.3, 7.8.

***Proof*** Suppose we are given any regular plane curve $\gamma: S^1 \to \mathbb{R}^2$. As stated above, applying Thom's Lemma to the map $\mu$ of 9.5 with $Q$ the submanifold of higher inflexions (resp. vertices) proves that a dense set of curves have only ordinary inflexions and vertices. Further taking $k = 3$ and $Q$ to be the $a_3$-axis (resp. $a_2$-axis) then the inflexions (resp. vertices) of $\psi \circ \gamma$ correspond to points $t$ with $\mu(t, \psi) \in Q$. However again by Thom's

Lemma we know that for a dense set of $\psi \in U_1$ the map $\mu(-, \psi) = v$ is transverse to $Q$. Consequently by 8.16(7) the set $v^{-1}(Q)$ of ordinary inflexions (resp. vertices) is finite for a dense set of $\psi \in U_1$, whence the result.

We now have to prove that these properties are open. We first show that this is so for the property of only having ordinary inflexions and vertices. Let $Q$ denote either the set $a_2 = a_3 = 0$ in $V_3$ (higher inflexions) or the set $a_3 = a_4 - a_2^3 = 0$ in $V_4$ (higher vertices, together with very high inflexions $a_2 = a_3 = a_4 = 0$). Thus $Q$ is closed in each case. Let $\tilde{\gamma}$: $S^1 \times U \to \mathbb{R}^2$ be a family of curves with $\tilde{\gamma}_0$ having only ordinary inflexions and vertices. Let $\mu : S^1 \times U \to V_k$ be the corresponding family of Monge–Taylor maps. Then the compactness of $S^1$, together with the fact that $\mu_0$ is transverse to $Q$ (misses $Q$ in fact), implies by 8.23 that $\mu(S^1 \times \{u\})$ misses $Q$ for $u$ in some open neighbourhood $U'$ of 0. Hence nearby curves $\tilde{\gamma}_u$ in the family also possess no higher inflexions and no higher vertices.

It remains to show that, in fact, the property of having *finitely many* ordinary inflexions and vertices and no higher ones is open. First note that, if $\gamma$ has an ordinary inflexion (resp. vertex) at $t \in S^1$, then the image of the map $\mu : S^1 \to V_3$ meets the $a_2$ (respectively $a_3$) axis at $\mu(t)$ and is transverse to this axis. For the method see 9.4(3).

Let $\tilde{\gamma} : S^1 \times U \to \mathbb{R}^2$ be a family of curves with $\tilde{\gamma}_0$ having finitely many ordinary inflexions and vertices, and no higher ones, so that $\mu(-, 0)$ is transverse to the $a_2$ and $a_3$-axes. Since transversality is an open condition when the source is compact and the relevant submanifold closed (8.23) it follows that $v = \mu(-, u) : S^1 \to V_3$ will also be transverse to these axes for all $u$ in some neighbourhood $U_1$ of $0 \in U$. Consequently, if $Q$ is either of these axes the set $v^{-1}(Q)$ of ordinary inflexions or vertices of $\tilde{\gamma}_u$ is finite and also there are no higher ones. (See 8.16(7).) This proves the result. (Actually one can prove the stronger assertion that for some neighbourhood $U_2 \subset U_1$ of 0 the curve $\tilde{\gamma}_u$ has *the same number* of vertices as $\tilde{\gamma}_0$.) □

We have shown that the properties in 9.7 are *stable*, while the contrary properties are *unstable*: deforming a curve in a finite dimensional family preserves the properties listed, while a curve not possessing some of these properties is arbitrarily close to curves which do possess all of them. That is the best we can do without a substantial increase in technicality. Perhaps one would like to say that a 'random curve' will have all the properties of 9.7 but there is no obvious way of making that idea precise. Certainly we expect to repeatedly observe only stable phenomena in nature (since experimental conditions vary) and the phenomena we have been studying *are* stable.

## 9.8    Exercises

(1) Using the Transversality Theorem 9.5 show that for any fixed $\kappa > 0$ any compact curve can be slightly perturbed in such a way that no vertex has curvature $\kappa$.

(2) Explain why, when $k = 3$, $\mu(I)$ can only touch the $a_3$-axis at the origin.

(3) (*i*) When studying the geometry of plane curves using distance-squared functions (resp. height functions) clearly the unit circle $C$ $(x^2 + y^2 = 1)$ (resp. the $x$-axis $L$) is a most degenerate curve. Find a 1-parameter family of polynomial maps $\psi : \mathbb{R}^2 \times (-1, 1) \to \mathbb{R}^2$, $\psi((x, y), s) = \psi_s(x, y)$ with $\psi_0 = \mathrm{id}$ and $\psi_s(C)$ (resp. $\psi_s(L)$) having only ordinary vertices (resp. ordinary inflexions) for $s \neq 0$. (Hint: keep the degree of $\psi$ small – very small for the circle.)

(*ii*) The deformation of $L$ you produced in (*i*) probably (indeed certainly) was not small. (For any $s \neq 0$ the map $\psi_s$ will move points of $L$ arbitrarily large distances.) Find a small *smooth* family of deformations $\psi_s$ of $L$ with the same properties, i.e. given any $\varepsilon > 0$ find such a $\psi$ with $\|\psi_s(x, 0) - (x, 0)\|$ $< \varepsilon$ for all $x \in \mathbb{R}$, $s \in (-1, 1)$. This time the deformations will not be polynomial.

## A transversality theorem for space curves

In this section we wish to prove results for space curves analogous to those obtained previously for plane curves. Again we shall need some way of storing the infinitesimal information concerning the local geometry of the curve at each point. One way of attempting to do this is to write the curve locally as the graph of a mapping, but this time with source the tangent line to the curve at the given point, and target the normal plane at the point (fig. 9.6). The infinitesimal information concerning the curve should then be stored in the $k$-jets of the components of this mapping. The main problem facing us should now be clear: how are we to choose coordinates in the normal plane? Each choice will yield two different

**Fig. 9.6.** Alternative coordinate axes in the normal plane

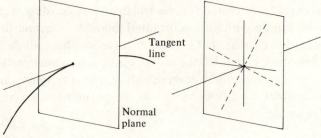

Tangent
line

Normal
plane

component functions. One natural choice would be to use the normal and binormal as coordinate axes, but unfortunately these are not always defined: consider for example a space curve containing some portion of a straight line. We can circumvent this problem as follows.

Let $\gamma: I \to \mathbb{R}^3$ be a (regular) space curve, with $I$ an open connected subset of the unit circle $S^1$, increasing $t$ corresponding to anticlockwise orientation of $S^1$. We now choose a smooth family of unit vectors $V(t)$, with $V(t)$ normal to $\gamma$ at $t$ so $\|V(t)\| = 1$ and $V(t) \cdot T(t) = 0$ for all $t \in I$. (Such 'vector fields' can be obtained as follows: consider the smooth map $T: I \to S^2$ which takes $t$ to the unit tangent vector $T(t)$. If $V$ is any vector in $S^2 - T(I)$ we can obtain the vector field $V(t)$ by orthogonally projecting $V$ onto each of the normal planes and normalising. Thus $V(t) = (V - (V \cdot T(t))T(t))$ $\|V - (V \cdot T(t))T(t)\|^{-1}$. Note that by Sard's Theorem there are plenty of vectors $V$ in the complement of the image of $T$.) We can now obtain a second smooth family of unit vectors $W(t) = T(t) \times V(t)$ (see 2.36(1)) normal to $\gamma$ at $t$. We now use the perpendicular lines spanned by $T(t)$, $V(t)$, $W(t)$ as axes at $\gamma(t)$ and the unit points on the axes corresponding to the three given vectors. Clearly at each point $\gamma(t)$ we can write $\gamma(I)$ locally as a graph $(\xi, f_t(\xi), g_t(\xi))$ with $f_t$ and $g_t$ smooth and $j^1 f_t(0) = j^1 g_t(0) = 0$. If, as before, $V_k$ denotes the space of polynomials in $\xi$ of degree $\geqslant 2$ and $\leqslant k$ we have a map, the *Monge–Taylor map for the space curve* $\gamma$, $\mu_\gamma: I \to V_k \times V_k$ given by $\mu_\gamma(t) = (j^k f_t(0), j^k g_t(0))$. Of course $\mu_\gamma$ depends rather heavily on our choice of unit normals $V(t)$.

One now sets about proving the transversality result in much the same way as before, using the space $P_k$ of polynomial maps $\mathbb{R}^3 \to \mathbb{R}^3$ of degree $\leqslant k$. The one significant change that we shall have to make concerns the vector field $V(t)$. If we deform the original curve what are we to replace $V(t)$ by? Since the deformation is obtained by composing with a polynomial map $\psi: \mathbb{R}^3 \to \mathbb{R}^3$, which is a diffeomorphism on some open set containing $\gamma(I)$, the vector $V(t)$ will be sent to some new vector $D\psi(\gamma(t))V(t)$ which will be neither zero nor tangent to $\psi \circ \gamma$ at $t$. Orthogonally projecting this new vector onto the normal plane to $\psi \circ \gamma$ at $t$ and normalising, we obtain the required new smooth family of normal vectors $V_\psi(t)$. So assuming as before that $I = S^1$, we choose an open neighbourhood $U$ of id $\in P_k$ consisting of polynomial maps which map an open set containing $\gamma(S^1)$ diffeomorphically to its image. We have now shown that there is a smooth map

$$\mu: S^1 \times U \to V_k \times V_k$$

defined by $\mu(-, \psi) = $ Monge–Taylor map for the curve $\psi \circ \gamma$ using the family of normal vectors $V_\psi(t)$.

**9.9    Theorem**

*The map $\mu: S^1 \times U_1 \to V_k \times V_k$ is a submersion, for some open neighbourhood $U_1 \subset U$ of* id.

**Proof** Again we need only consider a point $(t, \text{id}) \in S^1 \times P_k$ (compare 8.23, 9.5), and we may suppose that $\gamma(t)=(0, 0, 0)$, $T(t)=(1, 0, 0)$, $V(t)=(0, 1, 0)$ and $W(t)=(0, 0, 1)$. Locally then $\gamma(S^1)$ can be written in the form $\{(\xi, f(\xi), g(\xi))\}$, with $f(0)=g(0)=f'(0)=g'(0)=0$, so $\mu(t, \text{id})=(j^k f(0), j^k g(0))$.

Let $F(x)$, $G(x)$ be any polynomials in $x$ of degree $\geqslant 2$ and $\leqslant k$ and let $\psi^s(x, y, z)=(x, y+sF(x), z+sG(x))$. For $s$ small $\psi^s \in U$, $\psi^s(0)=0$ and $D\psi^s(0)=\text{id}$. So we must use the same $x, y, z$ coordinates at 0 for $\psi^s \circ \gamma$ as before, and since $\psi^s \circ \gamma(S^1)$ is given locally at $\psi^s(\gamma(t))=0$ by $\{(\xi, f(\xi)+sF(\xi), g(\xi)+sG(\xi))\}$ we have $\mu(t, \psi^s)=(j^k(f+sF)(0), j^k(g+sG)(0))$. So from the path $s \mapsto (t, \psi^s(x, y))$ in $S^1 \times U$ we obtain the corresponding tangent vector in $V_k \times V_k$ at $\mu(t, \text{id})$:

$$\lim_{s \to 0} (j^k(f+sF)(0)-j^k f(0), j^k(g+sG)(0)-j^k g(0))/s$$
$$=(j^k F(0), j^k G(0))=(F(\xi), G(\xi)).$$

Since $F$ and $G$ were any polynomials with $2 \leqslant \deg F$, $\deg G \leqslant k$ this clearly proves that $\mu$ is a submersion at $(t, \text{id})$ and the result follows.    $\square$

The applications of this result are given in the following three exercises.

**9.10    Exercises**

(1) Taking $Q$ to be the manifold $\{(0,0)\}$ in $V_2 \times V_2$ deduce that for a generic set of curves the curvature $\kappa$ never vanishes.

(2) Write down the conditions for the height function in the direction $(u, v, \sqrt{(1-u^2-v^2)})$ to have an $A_{\geqslant k}$ singularity at the origin when restricted to the curve $(\xi, a_2\xi^2+a_3\xi^3+\cdots, b_2\xi^2+b_3\xi^3+\cdots)$, for $1 \leqslant k \leqslant 4$. Verify that the curve has $A_{\geqslant 3}$ contact with some plane iff $a_2 b_3 - a_3 b_2 = 0$, and $A_{\geqslant 4}$ contact iff in addition $a_2 b_4 - a_4 b_2 = 0$. Away from the subspace $a_2 = b_2 = 0$ in $V_4 \times V_4$ the set $Q_1 = \{(a_2, \ldots, b_4): a_2 b_4 - a_4 b_2 = a_2 b_3 - a_3 b_2 = 0\}$ is a smooth manifold. Deduce from theorem 9.9 and Thom's Lemma that for a generic set of curves the height functions never have an $A_{\geqslant 4}$ (Use (1) above.) In particular it follows from 2.35 that if $\tau = 0$ at some point of one of these curves then $\tau' \neq 0$ at that point.

(3) Carry out the same procedure for the distance-squared function from the point $(u, v, w)$. In particular show that if the curve has $A_{\geqslant 4}$ contact with a sphere at 0 then either $a_2$ or $b_2 \neq 0$ and $\phi_1 = (a_2^2 + b_2^2)(a_3 b_2 - a_2 b_3) + (a_4 b_3 - a_3 b_4) = 0$. Provided $a_3$ and $b_3$ are not both zero we have $A_{\geqslant 5}$ contact iff $\phi_1 = \phi_2 = 2(a_2 a_3 + b_2 b_3)(a_3 b_2 - a_2 b_3) + (a_5 b_3 - a_3 b_5) = 0$. Prove that away from the subspace $Q_2$ given by $a_3 = b_3 = 0$ the set $Q_3$ in $V_5 \times V_5$ given by the vanishing of $\phi_1$ and $\phi_2$ is a smooth manifold. Use theorem 9.8

and Thom's Lemma (twice) to deduce that for a generic set of curves the distance squared functions never have an $A_{\geqslant 5}$. Prove further than generically these functions are always (p)versally unfolded in the family of all distance-squared functions.

# 10

## *More on unfoldings*

'What do you say, Watson?'
I shrugged my shoulders.
'I must confess that I am out of my depths,' said I.
(*The Stockbroker's Clerk*)

In this chapter we return to unfoldings of functions and give a proof of the main theorem, 6.6p, for *analytic* functions and families. The work involved in this is quite substantial, and a good deal more complicated than anything else in the book. There is, however, a relatively short initial section of the proof which at least makes the result quite plausible and is relatively easy to follow. Before plunging into the proof we give some explanation of why we shall only deal with the analytic case.

As we have mentioned before, the complete proof of the main theorem 6.6p for smooth functions and families unfortunately requires a rather formidable technical result called the Malgrange Preparation Theorem. This result, together with Sard's Theorem (4.18), forms the cornerstone of the theory of smooth maps, but a proof would be out of place in a book of this sort. (A complete proof, together with a proof of the main theorem on unfoldings, appears in Bröcker and Lander (1975).) What we do here is to assume that all our functions are analytic, that is given by convergent power series, and produce analytic families of functions which induce any unfolding of $t^{k+1}$ from the unfolding $G$ of 6.6p. This involves two steps: producing the power series and proving that they are convergent. The first is easy, the second harder and more technical – we expect the reader to have some familiarity with contour integration and the Maximum Modulus Principle from complex analysis. The proof we give is an adaptation of a more general one in Kas and Schlessinger (1972).

Although smooth functions have Taylor series we know (see 3.9, 3.10(2)) that a function and its Taylor series need have very little in common. So the result we prove is of no use in the smooth case. However there are two points in its favour (apart from the attraction of its being easier to prove!). First, the initial part of the proof, in addition to actually constructing the parametrized changes of variable needed to reduce any unfolding of $t^{k+1}$ to the standard one of 6.6p, leads in a rather natural way to the versality criterion of 6.10p. Moreover this part of the proof generalizes easily to the case of several variables $t_i$.

Secondly, once we have proved that the results of chapter 6 hold in the analytic case, the applications of chapter 7 all hold for analytic curves. Moreover the deformations used in chapter 9 to prove genericity results are all analytic (indeed polynomial), so we do obtain a complete treatment of the generic geometry of analytic plane and space curves.

**10.1    Theorem** (See 6.6p)

*The unfolding $G: \mathbb{R} \times \mathbb{R}^{k-1}, 0 \to \mathbb{R}$ defined by*

$$G(t, u) = \pm t^{k+1} + u_1 t + u_2 t^2 + \cdots + u_{k-1} t^{k-1}$$

*is a (p)versal unfolding of $g: \mathbb{R}, 0 \to \mathbb{R}, g(t) = \pm t^{k+1}$.*

What we show is that for any *analytic* unfolding $F: \mathbb{R} \times \mathbb{R}^r, 0 \to \mathbb{R}$ of $g$ we can find *analytic* maps $a: \mathbb{R} \times \mathbb{R}^r, 0 \to \mathbb{R}, 0, b: \mathbb{R}^r, 0 \to \mathbb{R}^{k-1}, 0$ and $c: \mathbb{R}^r, 0 \to \mathbb{R}, 0$ such that $a(t, 0) = t$ for all small $t$ and

$$F(a(t, x), x) = G(t, b(x)) + c(x). \tag{1}$$

This is the analytic version of 6.2p; the final step of the reduction, to 6.3p, proceeds as in chapter 6. We shall take $g(t) = t^{k+1}$ in what follows, in order to avoid a lot of $\pm$ signs.

## Determining the power series

First some notation. We shall write the power series we seek in the form

$$a(t, x) = a_0(t, x) + a_1(t, x) + \cdots + a_p(t, x) + \cdots$$
$$b(x) = b_0(x) + b_1(x) + \cdots + b_p(x) + \cdots$$
$$c(x) = c_0(x) + c_1(x) + \cdots + c_p(x) + \cdots$$

where $a_j, b_j, c_j$ are homogeneous polynomials of degree $j$ in $x_1, \ldots, x_r$. For $a$ the coefficients in these polynomials are power series in $t$. We also set

$$a^p = a_0 + \cdots + a_p, b^p = b_0 + \cdots + b_p, c^p = c_0 + \cdots + c_p.$$

Then (1) above is equivalent to the sequence of conditions

$$\left. \begin{array}{l} F(a^p(t, x), x) \equiv_{p+1} G(t, b^p(x)) + c^p(x) \\ a^p(t, 0) = t \end{array} \right\} \tag{1$_p$}$$

where $\equiv_{p+1}$ means that the two sides differ by terms of degree $\geq p+1$ in the variables $x_1, \ldots, x_r$.

Consequently we seek $a_j, b_j, c_j, 1 \leq j \leq p$, which solve (1)$_p$. The construction proceeds by induction on $p$.

To start the induction set $b_0(x)=0$, $c_0(x)=0$, $a_0(t, x)=t$. Then $(1)_0$ reduces to the fact that $F$ is an unfolding of $g$ at 0. Suppose that we have constructed $a^p(t, x)$, $b^p(x)$, $c^p(x)$ to satisfy $(1)_p$. Our object is to find $a_{p+1}(t, x)$, $b_{p+1}(x)$, $c_{p+1}(x)$ so that $(1)_{p+1}$ holds. This requires

$$F(a^p(t, x)+a_{p+1}(t, x), x)-(G(t, b^p(x)+b_{p+1}(x))+c^p(x)+c_{p+1}(x))$$

$$\equiv_{p+2} 0. \qquad (2)$$

By Taylor's theorem

$$F(u+v, x)=F(u, x)+v\frac{\partial F}{\partial t}(u, x)+\tfrac{1}{2}v^2\frac{\partial^2 F}{\partial t^2}(u, x)+\cdots$$

so that

$$F(a^p+a_{p+1}, x)=F(a^p, x)+a_{p+1}\frac{\partial F}{\partial t}(a^p, x)+F_1(t, x)$$

where all terms of $F_1$ have degree $\geqslant 2(p+1)\geqslant p+2$ in the $x$ variables. Since we are only interested in terms of degree $\leqslant p+1$ in the $x_i$ we can ignore $F_1$. Moreover $a_{p+1}$ is homogeneous of degree $p+1$ in $x$ so we need only consider the terms in $\partial F/\partial t(a^p, x)$ which do not involve $x$, in other words $\partial F/\partial t(a^p(t, 0), 0)$, which by $(1)_p$ is $\partial F/\partial t(t, 0)=(k+1)t^k$.

On the other hand, using the form of $G$,

$$G(t, b^p+b_{p+1})+c^p+c_{p+1}=G(t, b^p)+c^p+\left(\sum_{j=1}^{k-1}(b_{p+1})_j t^j+c_{p+1}\right)$$

where the subscript $j$ denotes the $j$th coordinate of $b_{p+1}(x)$. Consequently equation (2) above can be rewritten as

$$(F(a^p(t, x), x)-G(t, b^p(x))-c^p(x))+(k+1)t^k a_{p+1}(t, x)-\sum_{j=1}^{k-1}(b_{p+1}(x))_j t^j$$

$$-c_{p+1}(x)\equiv_{p+2}0. \qquad (3)$$

By $(1)_p$ the first bracketed expression in (3) contains no terms of degree $<p+1$ in $x$. So comparing coefficients of terms of degree $p+1$ the solution of (3) reduces to the problem of writing any power series $\theta(t)$ (obtained from a term of degree $p+1$ in the first bracketed expression) in the form

$$\theta(t)=(k+1)t^k\alpha(t)+\sum_{j=1}^{k-1}\beta_j t^j+\gamma \qquad (4)$$

for a suitable power series $\alpha(t)$ and constants $\beta_j$ and $\gamma$. Clearly any series *can* be represented in this way, in fact uniquely, from which it follows that $a_{p+1}$, $b_{p+1}$, $c_{p+1}$ are uniquely determined. This completes the inductive construction of the series $a$, $b$, $c$ and proves moreover that they are unique.

*Note* The same argument, done a little more carefully, shows that any unfolding $t^{k+1}+\sum_{i=1}^{k-1}u_i\phi_i(t)$ is (p)versal provided the $(k-1)$-jets of $\phi_1, \ldots,$

$\phi_{k-1}$ have coefficients which form a nonsingular $(k-1) \times (k-1)$ matrix – compare 6.10p. (That is to say the above argument produces power series $a, b, c$; one still has to prove convergence, as below.)

## Proof of convergence

We now consider the considerably more complicated problem of proving that the above series are in fact convergent. This clearly involves some estimate on the size of the terms $\alpha(t)$, $\beta_j$, $\gamma$ which arise in the solution of (3). Since (3) already involves $a^p, b^p, c^p$, any such estimates will necessarily be obtained by induction on $p$. To find the estimates we shall need to work with complex power series, but the above construction certainly produces real series from a real unfolding $F$ so we know that the $a, b, c$ will turn out to have real coefficients.

We need the following notation and technical results. All power series are to be expanded about the origin in some complex space $\mathbb{C}^m$.

### 10.2    Notation

(1) For $\varepsilon > 0$ let $D_\varepsilon$ denote the disc in the complex plane, $D_\varepsilon = \{t \in \mathbb{C} : |t| \leqslant \varepsilon\}$. For $m \geqslant 2$ the product $\{(x_1, \ldots, x_m) \in \mathbb{C}^m : |x_i| \leqslant \varepsilon, 1 \leqslant i \leqslant m\}$ is denoted by $B_\varepsilon$.

(2) Suppose that $h(t, x)$ (resp. $h(x)$) is analytic on some neighbourhood of 0 in $\mathbb{C} \times \mathbb{C}^r$ (resp. $\mathbb{C}^r$), and let $v = (i_1, \ldots, i_r)$ be a multi-index. Then $h_v(t)$ (resp. $h_v$) will denote the coefficient of $x_1^{i_1} \ldots x_r^{i_r}$ in the power series expansion of $h$. Further write $|v|$ for $i_1 + \ldots + i_r$.

(3) Let $h$ be analytic on some neighbourhood of $D_\varepsilon \times 0$ in $\mathbb{C} \times \mathbb{C}^r$ (resp. of 0 in $\mathbb{C}^r$) and let $g$ be analytic on some neighbourhood of 0 in $\mathbb{C}^r$. Then we write $h \ll g$ to mean $\sup_{t \in D_\varepsilon} |h_v(t)| \leqslant |g_v|$ (resp. $|h_v| \leqslant |g_v|$) for all $v$.

(4) For a vector valued analytic $h$, consisting of $k$ analytic functions, $h = (h_1, \ldots, h_k)$, we write $h \ll g$ when each $h_i \ll g$.

Our first lemma establishes certain useful properties of the symbol $\ll$, which are crucial in the comparison process by which we prove convergence.

### 10.3    Lemma

*Suppose $H, G, h_i, g_i, 1 \leqslant i \leqslant m$, are analytic functions, with*

$$H(x_1, \ldots, x_m) \ll G(x_1, \ldots, x_m),$$
$$h_i(y_1, \ldots, y_n) \ll g_i(y_1, \ldots, y_n) \qquad 1 \leqslant i \leqslant m$$

*and suppose that the coefficients in the power series expansions of $G$ and the*

$g_i$ *are all real and non-negative. Then*

$$H(h_1(y), \ldots, h_m(y)) \ll G(g_1(y), \ldots, g_m(y)). \tag{5}$$

**Proof** Write $H(x) = \sum H_\nu x^\nu$, $G(x) = \sum G_\nu x^\nu$, $h_i(y) = \sum h_{i\eta} y^\eta$, $g_i(y) = \sum g_{i\eta} y^\eta$, the summations being over all relevant multi-indices. We know that $|H_\nu| \leqslant G_\nu$, $|h_{i\eta}| \leqslant g_{i\eta}$, for all $\nu$, $i$, and $\eta$. We now wish to compare the coefficients of $y^\omega$ on the two sides of (5) for each multi-index $\omega$. There is clearly a polynomial $p_\omega$ in variables $(F_\nu, f_{i\eta})$ (where $\nu$ and $\eta$ take all values with $|\nu| \leqslant |\omega|$, $|\eta_i| \leqslant |\omega|$, $1 \leqslant i \leqslant m$) such that the coefficient of $y^\omega$ in the left side of (5) equals $p_\omega(H_\nu, h_{i\eta})$, while the coefficient of $y^\omega$ in the right side is $p_\omega(G_\nu, g_{i\eta})$. One easily verifies that the polynomial $p_\omega$ has non-negative coefficients.

Consequently the lemma will be proved if for any polynomial $p$ in $N$ variables $z_1, \ldots, z_N$ with non-negative coefficients and any numbers $c_1, \ldots, c_N$, $d_1, \ldots, d_N$ with $|c_i| \leqslant d_i$, $1 \leqslant i \leqslant N$ we have $|p(c_1, \ldots, c_N)| \leqslant p(d_1, \ldots, d_N)$. But clearly, writing $p(z_1, \ldots, z_N) = \sum p_\nu z^\nu$ we have

$$|p(c)| = \left|\sum p_\nu c^\nu\right| \leqslant \sum p_\nu |c^\nu| \leqslant \sum p_\nu d^\nu = p(d)$$

as required.    □

Taking $H(x_1, x_2) = G(x_1, x_2) = x_1 x_2$ we deduce:

### 10.4    Corollary

*If* $h_1 \ll g_1$ *and* $h_2 \ll g_2$ *and if the coefficients of* $g_1$ *and* $g_2$ *are real and non-negative, then* $h_1 h_2 \ll g_1 g_2$.    □

The next result we prove gives some estimate on the size of $\alpha(t)$, $\beta_j$, $\gamma$ appearing in the representation of $\theta(t)$ above in (4), in terms of $\theta$. This will clearly be crucial in obtaining our inductive estimate on the size of $a_p$, $b_p$, $c_p$. Write $\gamma$ as $\beta_0$.

### 10.5    Lemma

*Given any* $\varepsilon$ *with* $0 < \varepsilon < 1$ *there is a constant* $L > 0$ *such that every function* $\theta(t)$, *analytic on* $D_\varepsilon$, *can be written uniquely in the form*

$$\theta(t) = \alpha(t)(k+1)t^k + \sum_{j=0}^{k-1} \beta_j t^j$$

*with* $\alpha(t)$ *analytic on* $D_\varepsilon$, $|\beta_j| \leqslant LM$ *and* $\sup_{t \in D_\varepsilon} |\alpha(t)| \leqslant LM$, *where* $M = \sup_{t \in D_\varepsilon} |\theta(t)|$.

To establish this we need the following.

### 10.6    Cauchy's inequalities

*Let* $h$ *be a function analytic on* $B_\varepsilon \subset \mathbb{C}^n$, $h(x) = \sum h_\nu x^\nu$ *and suppose* $\sup_{x \in B_\varepsilon} |h(x)| \leqslant M$. *Then* $|h_\nu| \leqslant M\varepsilon^{-|\nu|}$.

**Proof** We first prove the case $n=1$. Here, if $h(x_1)=\sum_{p=0}^{\infty} h_p x_1^p$ then $h_p=(1/2\pi i)\int_{\gamma}(h(x_1)/x_1^{p+1})$, where this contour integral is over the path $\gamma(s)=\varepsilon e^{is}$. (This is a standard result from complex analysis; see for example Ahlfors (1966).)

Now

$$|h_p|=\frac{1}{2\pi}\left|\int_{\gamma}\frac{h(x_1)}{x_1^{p+1}}\right|\leqslant\frac{1}{2\pi}L(\gamma)\sup_{|x_1|=\varepsilon}\left|\frac{h(x_1)}{x_1^{p+1}}\right|\leqslant\varepsilon M\varepsilon^{-p-1}=M\varepsilon^{-p},$$

where $L(\gamma)$ is the length of the path $\gamma$, and this is the required result.

For the general case we proceed by induction on $n$. Writing $h(x)=\sum h_p(x_2,\ldots,x_n)x_1^p$, for any $x_2,\ldots,x_n$ with $|x_j|\leqslant\varepsilon$ for $2\leqslant j\leqslant n$, the case $n=1$ shows that $|h_p(x_2,\ldots,x_n)|\leqslant M\varepsilon^{-p}$. Applying the inductive hypothesis to $h_p$, for the same $M$ as in the statement of the lemma, now completes the induction. □

Using 10.6 we now establish 10.5.

Using the Taylor series of $\theta(t)$ it is clear that $\theta$ can be expressed uniquely in the form given in 10.5. Moreover from 10.6 it follows that $|\beta_j|\leqslant M\varepsilon^{-j}$ for $0\leqslant j\leqslant k-1$. However for $t\in D_\varepsilon$ we have

$$|(k+1)\theta(t)t^k|=\left|\theta(t)-\sum\beta_j t^j\right|\leqslant|\theta(t)|+\sum_{j=0}^{k-1}|\beta_j t^j|$$

$$\leqslant M+\sum_{j=0}^{k-1}M\varepsilon^{-j}\varepsilon^j=(k+1)M.$$

By the Maximum Modulus Principle (see for example Ahlfors (1966)) we have $\sup_{t\in D_\varepsilon}|(k+1)\theta(t)t^k|=(k+1)\varepsilon^k\sup_{t\in D_\varepsilon}|\theta(t)|$ so $\sup_{t\in D_\varepsilon}|(\theta(t)|\leqslant M\varepsilon^{-k}$ and taking $L=\varepsilon^{-k}$ the lemma is proved. □

The final ingredients of the proof of convergence are two convergent power series which will be used in the comparison process. Let $e$, $d$, $E$, $D$ be real and positive and set

$$A(x)=\frac{e}{4d}\sum_{j=1}^{\infty}\frac{d^j}{j^2}(x_1+\cdots+x_r)^j,$$

$$B(s,x)=\frac{E}{D}\sum_{j=1}^{\infty}D^j(s+x_1+\cdots+x_r)^j.$$

The power series $A$ and $B$ have the following important properties.

## 10.7 Lemma

(i) $A$ (resp. $B$) *is convergent in some neighbourhood of* 0 *in* $\mathbb{C}^r$ (resp. $\mathbb{C}\times\mathbb{C}^r$).

(ii) $(A(x))^p\ll(3e/d)^{p-1}A(x)$ *for* $p\geqslant 2$.

***Proof*** (i) Using the ratio test it is easy to see that the power series $\sum_{j=1}^{\infty} D^j t^j$, $\sum_{j=1}^{\infty} (d^j/j^2)t^j$ are analytic at $t=0 \in \mathbb{C}$ with radius of convergence $1/D$, $1/d$ respectively. Clearly $A$ and $B$, being the composites of analytic functions, are analytic. (Indeed $A$ (resp. $B$) is convergent in the interior of the closed set $B_{1/dr}$ (resp. $B_{1/D(r+1)}$).)

(ii) Set $A_1(t)=e/4d\sum (dt)^j/j^2$; then

$$(A_1(t))^2 = \left(\frac{e}{4d}\right)^2 \sum \lambda_j(dt)^j \quad \text{where} \quad \lambda_j = \sum_{p=1}^{j-1} 1/p^2(j-p)^2.$$

Writing $c_v$ for the binomial coefficient of $x^v = x_1^{i_1} x_2^{i_2} \cdots x_r^{i_r}$ in $(x_1 + \cdots + x_r)^{|v|}$, the coefficient of $x^v$ in $(A(x))^2$ is $(e/4d)^2 c_v \lambda_{|v|} d^{|v|}$. The proof of (ii) now proceeds by induction on $p$, the hardest step being the first. To establish the case $p=2$ we need to show that

$$(e/4d)^2 c_v \lambda_{|v|} d^{|v|} \leqslant (e/d)(3e/4d)c_v d^{|v|}|v|^{-2},$$

i.e. $\lambda_{|v|} \leqslant 12|v|^{-2}$ or $\lambda_j \leqslant 12/j^2$ for all $j \geqslant 2$. But this is a straightforward exercise in calculus.

We may now invoke 10.4 to provide the inductive step. For assuming by induction $(A(x))^p \ll (3e/d)^{p-1}A(x)$ we have by 10.4

$$(A(x))^{p+1} = (A(x))^p A(x) \ll (3e/d)^{p-1}A(x)A(x)$$

which from the case $p=2$ is $\ll (3e/d)^p A(x)$ as required.    □

Finally we come to the proof of convergence. This is done in two stages: in the first stage we set up the first step of the induction and prove the key result needed for the induction step. These are in 10.8 below. In the second stage (10.9) we obtain the comparison results necessary for the proof of convergence. We suppose now that $F(t, x)$ is analytic in a neighbourhood of $D_{2\varepsilon} \times B_\varepsilon \subset \mathbb{C} \times \mathbb{C}^r$.

## 10.8    First main lemma

*We may choose $D$, $E$, $e > 0$ so that*

(i) $a^1(t, x) - t$, $b^1(x)$, $c^1$, $x_i$ *are all* $\ll A(x)$ $(1 \leqslant i \leqslant r)$,

(ii) $F(t+s, x) - F(t, 0) \ll B(s, x)$.

***Proof*** (i) Writing $a^1(t, x) - t = \sum_{j=1}^{r}(a_1(t))_j x_j$, $b^1(x) = \sum_{j=1}^{r}(b_1)_j x_j$, we choose $e$ so that $e/4 \geqslant \max \{\sup_{t \in D_\varepsilon}|(a_1(t))_j|, |(b_1)_j|, |c_1|, 1\}$. Note that this choice is independent of $d$.

(ii) Let $M$ be the supremum over $(t, s, x) \in D_\varepsilon \times D_\varepsilon \times B_\varepsilon$ of $|F(t+s, x) - F(t, 0)|$, and write, for $|t| \leqslant \varepsilon$,

$$F(t+s, x) - F(t, 0) = \sum g_{pv}(t)s^p x^v.$$

By Cauchy's inequalities 10.6 we have $|g_{pv}(t)| \leqslant M\varepsilon^{-(p+|v|)}$ for all $t \in D_\varepsilon$. Writing $c_{p,v}$ for the binomial coefficient of $s^p x^v$ in $(s+x_1 + \cdots + x_r)^{p+|v|}$, the coefficient of $s^p x^v$ in $B(s, x)$ is $ED^{p+|v|-1}c_{p,v}$. Choosing $D \geqslant 1/\varepsilon$, $E \geqslant M/\varepsilon$

and using $c_{p,v} \geqslant 1$ we have

$$|ED^{p+|v|-1}c_{p,v}| \geqslant ED^{p+|v|-1} \geqslant M\varepsilon^{-(p+|v|)} \geqslant |g_{pv}(t)|$$

for all $t \in D_\varepsilon$, as required. $\qquad\qquad\qquad\qquad\qquad\square$

The point of (*i*) above is that it starts an inductive estimate on the size of $a_p(t, x)$, $b_p(x)$, $c_p(x)$. 10.8(*ii*) provides the induction step (compare equation (3) above). Write the formal power series $a$, $b$, $c$ obtained above in the form

$$a(t, x) - t = \sum a_v(t)x^v, \quad b(x) = \sum b_v x^v, \quad c(x) = \sum c_v x^v.$$

## 10.9      Second main lemma

*For a suitable choice of d we have*

$$a_v(t)x^v, b_v x^v, c_v x^v \quad all \ll A(x) \quad for \ all \ v \ with \ |v| \geqslant 1.$$

*In other words, $a^p(t, x) - t$, $b^p(x)$, $c^p(x)$ are all $\ll A(x)$ for all p.*

**Proof** The case $|v| = 1$ is dealt with by the proof of 10.8(*i*). Before starting the induction we note that, denoting the linear part in $s$ and $x$ of $F(t+s, x) - F(t, 0)$ by $\mathscr{L}(t, s, x)$, we have, by 10.8(*ii*),

$$F(t+s, x) - F(t, 0) - \mathscr{L}(t, s, x) \ll (E/D) \sum_{j=2}^{\infty} D^j(s+x_1 + \cdots + x_r)^j. \quad (6)$$

Now let $p \geqslant 1$ and suppose the lemma is proved for $|v| \leqslant p$. In what follows the subscript $v$ denotes the coefficient of $x^v$ in the indicated expression, as before.

Now with $|v| = p + 1$,

$$F(a^p(t, x), x)_v = (F(t + (a^p(t, x) - t), x) - F(t, 0) - \mathscr{L}(t, a^p(t, x) - t, x))_v.$$

By (6) above, 10.3 and the inductive hypotheses $a^p(t, x) - t \ll A(x)$, $x_i \ll A(x)$ we find that

$$F(a^p(t, x), x)_v \ll (E/D) \sum_{j=2}^{\infty} D^j((r+1)A(x))^j.$$

Using $(A(x))^j \ll (3e/d)^{j-1} A(x)$ this implies that

$$F(a^p(t, x), x)_v \ll (E/D) \left( \sum_{j=2}^{\infty} D^j(r+1)^j(3e/d)^{j-1} A(x) \right).$$

Provided $0 < 3D(r+1)e/d < 1$, we can sum this geometric series and obtain $3ED(r+1)^2 eA(x)/(d - 3D(r+1)e)$. On the other hand since $G(t, b^p(x)) = t^{k+1} + \sum_{j=1}^{k-1} (b^p(x))_j t^j$ we obtain $G(t, b^p(x))_v = 0$. Hence

$$(F(a^p(t, x), x) - G(t, b^p(x)))_v \ll 3ED(r+1)^2 eA(x)/(d - 3D(r+1)e).$$

We may now invoke 10.5 to deduce that the terms $a_v(t)$, $b_v$, $c_v$ obtained by solving equation (3) above (namely the $\alpha(t)$, $\beta_j$ and $\gamma$ of (4)) satisfy

$$\sup_{t \in D_\varepsilon} |a_v(t)|, |b_v|, |c_v| \text{ all } \ll 3LED(r+1)^2 eA_v/(d - 3D(r+1)e),$$

where $L$ is independent of $v$.

Now choose $d$ so large that $0 < 3D(r+1)e/d < 1$ and $0 < 3ED(r+1)^2e/(d-3D(r+1)e) < 1/L$, this choice being independent of $v$. We now deduce that

$$\sup_{t \in D_\varepsilon} |a_v(t)|, |b_v|, |c_v| \text{ all } \ll A_v, \text{ in other words}$$

$$a^p(t, x)-t, b^p(x), c^p(x) \text{ all } \ll A(x) \text{ for all } p. \qquad \square$$

We can now deduce that the power series $a$, $b$, $c$ obtained in the first half of the proof of 10.1 are convergent. For

$$a(t, x)-t = \lim_{p \to \infty} (a^p(t, x)-t) \ll A(x)$$

$$b(x) = \lim_{p \to \infty} b^p(x) \ll A(x)$$

and similarly $c(x) \ll A(x)$. Since $A(x)$ is convergent in some neighbourhood of the origin it easily follows from the comparison test that $a$, $b$ and $c$ are also analytic in some neighbourhood of the origin. $\qquad \square$

### 10.10    Exercise

Establish the inequality $\lambda_j \leqslant 12/j^2$ used in the proof of 10.7, as follows. Write $f(x) = 1/x^2(j-x)^2$, so that

$$\lambda_j = f(1)+f(2)+ \cdots +f(j-1) \quad \text{and} \quad f(x)=f(j-x).$$

(i) Show that $f$ is decreasing for $1 \leqslant x \leqslant j/2$ and increasing for $j/2 \leqslant x \leqslant j-1$.

(ii) Splitting into cases $j$ even, $j$ odd, show that, in order to prove $\lambda_j \leqslant 12/j^2$, it is enough to prove

$$\int_1^{j-1} f(x)\,dx + 2f(1) \leqslant 12/j^2.$$

(iii) Evaluating the integral, show that it is $\leqslant 4/j^2$ for $j \geqslant 2$. Now use the value of $f(1)$.

# 11

## *Surfaces, simple singularities and catastrophes*

'The giant rat of Sumatra, a story for which the
world is not yet prepared.' (*The Sussex Vampire*)

Having investigated, in such a thorough manner, the geometry of curves, and in particular (or, rather, in general!) generic curves, it is natural to enquire about the geometry of surfaces. Will the same methods yield similar results? In short the answer is yes. Indeed the only aspect of our investigation which requires a significantly different approach is the corresponding classification of functions. The theory of functions of two variables is decidedly more complicated (or, more positively, is *richer*) than that of functions of a single variable, as we shall now see.

### Functions of two variables

There is a natural generalization of our definition of right equivalence of functions of one variable to the many-variable case. Let $U_i$, $i = 1$, $2$ be open subsets of $\mathbb{R}^n$, and let $t_i$ be a point of $U_i$ with $f_i: U_i \to \mathbb{R}$ smooth functions.

### 11.1    Definition

We say that $f_1$ (at $t_1$) and $f_2$ (at $t_2$) are *right equivalent* (written $\mathcal{R}$-equivalent) if there exist open neighbourhoods of $t_i$, say $V_i \subset U_i$, a diffeomorphism $h: V_1 \to V_2$ and a constant $c \in \mathbb{R}$ such that

$$h(t_1) = t_2 \text{ and } f_1(t) = f_2(h(t)) + c$$

for all $t \in V_1$. (So $f_1$ is obtained from $f_2$ by a change of coordinates and addition of a constant.)

Given a smooth function $f_1: U_1 \to \mathbb{R}$ as above we can consider the level set $f_1^{-1}(f_1(t_1))$. For functions of one variable these level sets are clearly rather uninteresting! But for functions of two or more variables they can be quite complicated. Moreover it is immediate from 11.1 that right equivalent functions have locally diffeomorphic level sets. (The map $h$ above gives a diffeomorphism $V_1 \to V_2$ taking $f_1^{-1}(f_1(t_1))$ to $f_2^{-1}(f_2(t_2))$.) We now consider some examples.

## 11.2    Examples

(1) The natural generalizations of the $A_k$ singularities of chapter 3 to functions of *two* variables are the functions $f(t_1, t_2) = \varepsilon_1 t_1^{k+1} + \varepsilon_2 t_2^2$, defined in a neighbourhood of $(0, 0)$, where $\varepsilon_1, \varepsilon_2$ are $\pm 1$. The zero level sets are sketched in fig. 11.1, where the cusp gets tighter and the contact between curves gets greater as $k$ increases.

The next example may appear rather innocuous. Do not be misled! It is in fact:

### (2)  *The giant rat of Sumatra*

Consider the family of functions $f_a(t_1, t_2) = t_1 t_2 (t_1 - t_2)(t_1 - a t_2)$ $(a(a-1) \neq 0)$ at $(t_1, t_2) = (0, 0)$. The zero level sets of these functions consist of four distinct lines through the origin (fig. 11.2).

Suppose two of these functions $f_a$ and $f_b$ are right equivalent. Then there is a diffeomorphism $h$ taking some neighbourhood $V_1$ of $(0, 0)$ to another such neighbourhood $V_2$ preserving the level sets. Clearly $h$ must take each line in the zero set of $f_a$ to one of the lines in the zero set of $f_b$, so $Dh(0)$ will do the same. Those of you familiar with a little projective geometry will

**Fig. 11.1.** The sets $f(t_1, t_2) = 0$

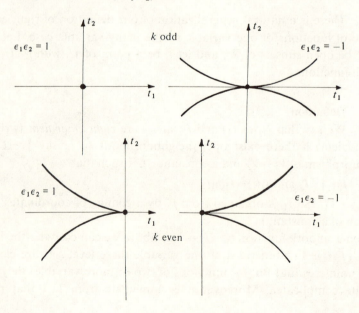

**Fig. 11.2.** The Giant Rat of Sumatra

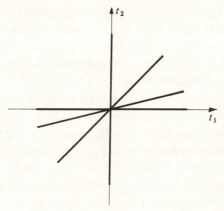

recall however that generally there is *no linear map* taking four concurrent lines to another such four lines. In fact it is a straightforward exercise (exercise 11.3(1) to be exact) to show that the four lines $f_a = 0$ can be mapped to the four lines $f_b = 0$ via a linear transformation if and only if $b = a$, $1/a$, $1-a$, $1/(1-a)$, $1-1/a$ or $1-1/(1-a) = -a/(1-a)$.

Indeed if we define $j(x)$ to be the expression $(x^2 - x + 1)^3/x^2(x-1)^2$ the zero sets of $f_a$ and $f_b$ are diffeomorphic if and only if $j(a) = j(b)$ (see 11.3(2)). The function $j$ then distinguishes right equivalent functions, so this relatively simple family of functions has a *modulus*, that is has uncountably many distinct types. This is in sharp contrast to our results of chapter 3 where we showed that there are *essentially* only countably many distinct types of singularities of functions of one variable. (One has to exclude the flat functions, but these really can be safely ignored.)

This occurrence of a modulus is rather more than inconvenient in any attempt to generalize the methods of this book. It is absolutely disastrous. Consider briefly the key steps in our applications of singularity theory to geometry. They are (*i*) list singularities; (*ii*) prove uniqueness of unfoldings; (*iii*) obtain models for the geometry of the versal unfolding; (*iv*) prove a transversality theorem to show that all functions of the family that describes the geometrical phenomenon we are studying are versally unfolded by the family; (*v*) use the models obtained in (*iii*) to describe some interesting geometry.

One obvious problem a modulus presents is to step (*iii*). An uncountable number of distinct functions lead to an uncountable number of unfoldings and hence an uncountable number of models. This situation is just about as bad as having no models at all. On the other hand, we claim that countably many models, such as those provided by the $A_k$ singularities, *is* an acceptable situation. Countable infinity can be visualized using the same imagination with which we visualize large numbers*. Discreteness is clearly the key. Perhaps more importantly however, for functions of one variable, in each geometrical application we need only consider finitely many of the $A_k$ singularities. (Recall 8.20(5).)

Even worse news is that step (*iv*) in the programme breaks down when a modulus appears. This breakdown is not due to any lack of ingenuity in proving the required transversality result. It is the fundamental problem, encountered in 8.20(2), that one cannot ensure that uncountably many transversality conditions will hold. Perhaps the best way to describe the problem here is by considering the method by which we proved our trans-versality results in chapter 9. Suppose (use your imagination!) that to describe the contact between a generic plane curve and circles one had to consider uncountably many singularities $X_t$, $t \in (0, 1)$, instead of simply $A_1, A_2$ and $A_3$. Consider the following property: each of these singularities, when it occurs, is versally unfolded by the family of distance-squared functions. It is reasonable to hope that this property is generic. In that case the condition for $X_t$ to occur and to fail to be versally unfolded might determine a space curve $C_t$ in the space $V_4$ (see 9.1), which generically the Monge–Taylor map $\mu$ would miss. (Compare the case of the distance-squared functions on a space curve, and the condition that an $A_4$ singularity is *not* versally unfolded, in 7.6(3) *and* 9.10(3).) However, the union of the curves $C_t$, $t \in (0, 1)$, would then trace out a *surface* (fig. 11.3), and we cer-

**Fig. 11.3.**

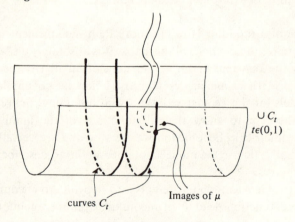

$\cup C_t$
$t \in (0,1)$

curves $C_t$

Images of $\mu$

---

* Psychologists assure us these begin at around 7.

tainly could not expect the image of $M$ to generically miss this surface. Thus it seems plausible that step (*iv*) may cause problems and indeed it does.

How do we overcome these difficulties? One way is to consider only those functions which do *not* possess such moduli. (The all too frequent device of defining one's way around a problem.) One might be a little shamefaced about this approach were it not for the two following remarks. First, it turns out that these so called *simple* singularities or functions are one of the most interesting geometrical finds of recent years. More importantly perhaps, from a practical point of view, they are also the only relevant functions when one is considering families of functions with fewer than six parameters. In particular they suffice to describe those geometrical phenomena arising in ordinary Euclidean 3-space that *can* be described using functions. It is these simple functions that we will consider in the next section.

Before passing on however we must briefly mention another, more honest, attempt to deal with the many problems caused by the appearance of moduli. We have seen that the family $f = t_1 t_2 (t_1 - t_2)(t_1 - a t_2)$ gives uncountably many inequivalent types (at least say for $0 < a < \frac{1}{2}$, see 11.3(1).) Yet this inequivalence is a subtle thing; the difference between distinct types is hardly apparent to the eye. Perhaps right equivalence is not the right equivalence to work with! Perhaps we should weaken the condition of 11.1 that $h$ is a local diffeomorphism to '$h$ is a local homeomorphism'. In fact one *can* weaken our definition of right equivalence and regain the paradise from which we fell (at the hands (or paws) of the giant rat of Sumatra, and his even more horrific brethren). One does not in fact weaken the definition 11.1 as suggested above: topological equivalence is *too* weak to provide a workable theory (or even the right answers, see 11.3(4)). Instead one has to work with 'universal stratified equivalence', which is far too technical for us to discuss here. The corresponding theory is due to Eduard Looijenga.

The cynics amongst you must now surely ask the price one must pay for regaining paradise in this manner, and there is a price to pay. The price of computability. The methods involved in setting up the above theory are absolutely useless when it comes to exhibiting the models we require. Even for the giant rat of Sumatra one has little idea what these models are. The world is indeed not yet prepared for its story. One should not construe this as a criticism of Looijenga's results. The truth of the matter is that these models will be extremely complicated, and no amount of general theory will circumvent the extremely difficult problem of describing them, requiring a great deal of detailed analysis.

## 11.3   Exercises

(1) If $L : \mathbb{R}^2 \to \mathbb{R}^2$ is a linear map taking the four lines $t_1 t_2 (t_1 - t_2)(t_1 - a t_2) = 0$ to $t_1 t_2 (t_1 - t_2)(t_1 - b t_2) = 0$ convince yourself that $b = a$, $1/a$, $1 - a$, $1/(1 - a)$, $1 - 1/a$ or $1 - 1/(1 - a)$. To actually prove this is so requires going through

a rather exhausting number (24) of possibilities. Go through some of the cases at least, and prove that if $b$ does take one of the four values above there is a linear map $L$ with the required properties. Can you identify the *group* at work here?

Deduce that, for $0 < a < \frac{1}{2}$, all the functions $f_a$ are inequivalent.

(2) Let $j(x) = \dfrac{(x^2 - x + 1)^3}{x^2 (x - 1)^2}$     $(x \neq 0, 1)$.

Check that, if $a$ and $b$ are related as in (1), then $j(a) = j(b)$. (Why is it enough to check this with $b = 1/a$ and $b = 1 - a$?) It actually follows that *if $j(a) = j(b)$, then $a$ and $b$ are related as in (1)*; can you see why? (Strictly this uses the fact that real polynomials in two variables $a$ and $b$ form a unique factorization domain.)

(3) Draw the zero level sets of the functions $f : \mathbb{R}^2 \to \mathbb{R}$ defined by $f = t_1^2 t_2 \pm t_2^{k-1}$, $t_1^3 \pm t_2^4$, $t_1^3 + t_1 t_2^3$, $t_1^3 + t_2^5$.

(4) If in definition 11.1 we replace 'diffeomorphism' by 'homeomorphism' show that $f_1(t_1) = t_1^p$ (resp. $f_1(t_1, t_2) = t_1^p \pm t_2^q$) is equivalent to $f_2(t_1) = t_1$ (resp. $f_2(t_1, t_2) = t_1$) if $p$ is odd (resp. $p$ or $q$ is odd).

## Simple singularities

'You have degraded what should have been a course of lectures into a series of tales'     (*The Copper Beeches*)

To attach a single name to the discovery of the simple singularities is impossible. These functions have arisen in the work of a great many mathematicians, for example F. Klein, P. Du Val, D. Kirby, M. Artin, E. Brieskorn, G. N. Tjurina, V. I. Arnold. However, if one views these objects as we have, as the functions that do not have moduli, then the relevant name is that of the Russian mathematician, Vladimir Arnold, who has, amongst many notable achievements, published a great deal of profound work in this area. (His papers feature in our list of recommended further reading.)

For functions of two variables the simple singularities consist of all functions right equivalent to one of the following.

I    $A_k$    $f : \mathbb{R}^2, 0 \to \mathbb{R}, 0$

         $f(t_1, t_2) = \pm t_1^{k+1} \pm t_2^2$ for $k \geqslant 1$

II    $D_k^{\pm}$    $f : \mathbb{R}^2, 0 \to \mathbb{R}, 0$

         $f(t_1, t_2) = t_1^2 t_2 \pm t_2^{k-1}$ for $k \geqslant 4$

III      $f : \mathbb{R}^2, 0 \to \mathbb{R}, 0$

   $E_6^{\pm}$    $f(t_1, t_2) = t_1^3 \pm t_2^4$

   $E_7$    $f(t_1, t_2) = t_1^3 + t_1 t_2^3$

   $E_8$    $f(t_1, t_2) = t_1^3 + t_2^5$.

The labels $A_k$, $D_k$, $E_6$, $E_7$, $E_8$ reflect a connection between these functions and certain Lie algebras which were already labelled in this way. Note that we have used the same label for $f(t_1, t_2) = \pm t_1^{k+1} \pm t_2^2$ as we have for our familiar $f(t_1) = \pm t_1^{k+1}$. The reason for this is that an addition of a square $\pm t_2^2$ essentially does not alter the geometry of the discriminant or bifurcation sets. Indeed, by adding squares to the above list one obtains all simple singularities in any number of variables.

We mentioned that the simple singularities are those which arise in describing geometric phenomena in ordinary Euclidean space. Indeed if we study generic surfaces by using the family of height functions (respectively distance squared functions), we need only consider $A_1$, $A_2$, $A_3$ (respectively $A_1$, $A_2$, $A_3$, $A_4$, $D_4$) singularities.

The height functions and their singularities measure the contact of the surface with its tangent planes. For example given a compact surface $M$ in Euclidean space one has a Gauss map $G: M \rightarrow S^2$ where $G(p)$ is the outward unit normal to $M$ at $p$ (perpendicular to the tangent plane $M_p$). Generically this map is a local diffeomorphism at each point of $M$ which has $A_1$ contact with its tangent plane. At points $p$ of a smooth curve the surface has $A_2$ contact with its tangent plane, and the Gauss map has a fold at each point along this curve (compare 7.13). ($A_2$ contact corresponds to the Gaussian curvature of $M$ vanishing; see fig. 11.4). At special points of this curve the Gauss map has a cusp (7.13), which corresponds to the surface having $A_3$ contact with its tangent plane.

The pedal of a surface is defined as for curves, and the pedal surface is rather like a Euclidean dual. As in the space curve case (see 7.9) the $A_2$ singularities give rise to a cuspidal edge on the pedal, while the $A_3$ points give swallowtails.

The distance-squared functions measure the contact of the surface with spheres. In the case of plane curves there is a single circle which has $A_{\geqslant 2}$ contact with the curve at each point. For a surface generally there are two spheres having $A_{\geqslant 2}$ contact with the surface at each point; these are

**Fig. 11.4.** Contact with tangent planes

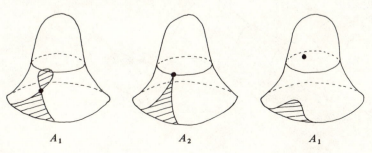

$A_1$            $A_2$            $A_1$

the spheres of curvature. This contact can generically be of type $A_3$ and $A_4$. The locus of the centres of these spheres of curvature is called the *focal set*. It is the envelope, in a certain sense, of the normals to $M$. Corresponding to the curve of $A_3$ singularities we have a cuspidal edge on the focal set, while an $A_4$ yields a swallowtail.

The new models in the surface case arise at *umbilic points* of the surface, when the two spheres of curvature at a point coincide. Hence the two sheets of the focal set come together at the corresponding unique centre of curvature. Generically there are two types of umbilic determined by the number of $A_3$ curves through the point in question. When there is only one such line we have a hyperbolic umbilic; when there are three we have an *elliptic umbilic*. The *hyperbolic umbilic* corresponds to the singularity $D_4^+ (f = t_1^2 t_2 + t_2^3)$, the elliptic umbilic to $D_4^- (f = t_1^2 t_2 - t_2^3)$. The corresponding focal sets are sketched in fig. 11.5.

## Catastrophes

'What is this, Holmes?' I cried.
This is beyond anything which I
could have imagined.'
(*The Resident Patient*)

Arnold's classification of the simple singularities was an extension of René Thom's famous earlier list of 'elementary catastrophes'. These seven singularities coincide with the functions $A_2$, $A_3$, $A_4$, $A_5$, $D_4^+$, $D_4^-$ and $D_5$ of Arnold. Thom had rather more picturesque titles for them, namely $A_2$ = 'fold', $A_3$ = 'cusp', $A_4$ = 'swallowtail', $A_5$ = 'butterfly', $D_4^+$ = 'elliptic umbilic', $D_4^-$ = 'hyperbolic umbilic', $D_5$ = 'parabolic umbilic'. (The umbilics are so labelled because of their connexion with umbilics on surfaces described in the last section.)

**Fig. 11.5.** Bifurcation sets

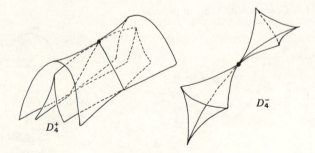

In a few pages we can hardly hope to explain or do justice to catastrophe theory (CT), and we refer the reader to the books Poston and Stewart (1978) and Zeeman (1977). (See in particular Zeeman's excellent short survey article in *Scientific American*, reprinted in his book, which gives a good flavour of the subject's methods and applications.) Nevertheless we feel that we should attempt to explain the connection between CT and the mathematics discussed in this book.

Let us return to chapter 1 and Poston's gravitational catastrophe machine. Here we have a family of potential functions (potential energy due to gravity in fact) determining the equilibrium position of a plane lamina. The family is parametrized by the position of the centre of gravity of the lamina, and equilibrium positions correspond to minima of the potential function. Furthermore the whole family is a (p)versal unfolding of each individual potential function (see 7.4(2)).

This situation, where the state of some system is determined by a potential function, is very common. It occurs throughout mechanics, where the total energy of the system seeks to be minimized, and also in geometric optics. (By Fermat's principle the path traversed by a ray of light is extremal with respect to the time taken to traverse it among paths with the same end points. Here we are seeking *both* maxima and minima of the relevant time function. Compare 2.28(1).)

(Note that the $D_4^{\pm}$ and $D_5$ singularities will need at least two state variables to be present before they can arise. Compare the discussion of distance-squared functions in the previous section.)

Now (p)versal unfoldings can be defined as for functions of one variable, and 'in general' the whole family of potential functions will be a (p)versal unfolding of each singularity of a function in the family. One of the most surprising aspects of Thom's results is that, provided one is interested in the way in which the stable equilibria change and jump as the control variables are altered, the *number* of state variables is irrelevant. What *is* relevant is the number of control variables. With one control variable this restricts the singularities to folds, with two to folds and cusps, with three to folds, cusps, swallowtails and elliptic and hyperbolic umbilics, and with four to all these and butterflies and parabolic umbilics. Thom stopped at four control variables on the grounds that these will be enough if the control space is ordinary space time. (In the introductory example of chapter 1 the control space was the plane.) Subsequently some investigations, notably those of M. Berry in optics, have used many more control variables.

The significance of Thom's idea lies in the hope that extremely complicated systems, such as those occurring in biology, where the number of state variables may be immense (say of the order of $10^4$, describing the concentrations of various chemicals throughout a cell), might nevertheless exhibit one of the standard catastrophe geometries (part of this geometry is the bifurcation set). Clearly one can only hope to model such phenomena implicitly.

The key underlying assumption, of course, is that 'in general' the families will versally unfold their individual potential functions and hence, because we are only considering a small number of control or unfolding parameters, there will only be a finite number of catastrophe geometries to observe. This idea is justified on the grounds that versality can be interpreted geometrically as a transversality condition which we expect to hold generally. (Compare 6.21, 8.20(5), (6).) Moreover, the reasoning continues, a non-versal unfolding is a transient and unstable state and so will not in practice be observed. This underlying assumption will not be valid in certain situations, where, for example, some inbuilt symmetry may invalidate the appeal to the typicality of transversality. (Thom, provocative as ever, has even suggested that it is these situations, where elementary catastrophe theory does *not* apply, which are the interesting ones!). Even here models have been constructed, much influenced by Thom's ideas, which have proved of great value.

In other situations the initial models suggested by CT have been replaced by more accurate and sophisticated ones, but again often very much influenced by the original. An example is the work of Golubitsky and Schaeffer (1979), which improves Zeeman's discussion of Euler Buckling. (One should bear in mind here that the perfect model is no model at all, simply the original object.)

Perhaps the most controversial applications are in the social sciences. (There appears to be a general, and understandable, dislike of any mathematical models at all in this area, which naturally raise the spectre of sinister social engineering.) The models are of course as vague as any which occur in the social sciences. But CT *does* provide models, of which the social scientist should be aware. (Just as an economist who believes all functional relationships are linear, should be made aware of exponential growth.) One might view the folded surface and cusp of the cusp catastrophe as a sophisticated proverb, which will often prove useful in describing otherwise indescribable phenomena.

In this book we have dealt exclusively with applications of CT to geometry. There can be no objections to its use here since the families of functions are exhibited explicitly and one can check the transversality conditions underlying the use of the CT models. It was Thom who originally suggested the use of CT in the study of the geometry of curves and surfaces in Euclidean space in his book (1975). I. R. Porteous described the associated geometry in a paper (1971). E. Looijenga then proved the relevant transversality results in 1974 (this forms a small part of his thesis, devoted largely to solving the problem of moduli described in the first section of this chapter). An excellent reference on the applications of CT, and singularity theory in general, to geometry is Wall's survey article (1976), although the reader should be warned that a good deal of work has been done since then; see especially Arnold (1981), Banchoff, Gaffney, McCrory (1982).

In its applications within geometry then CT has been and is being extremely successful. Having something new, and very relevant, to say about such a well-worked subject as the geometry of curves and surfaces is an accomplishment indeed. Surely such a brilliant circle of ideas deserves applause from all.

## 11.4    Project

'Well, really, I came to seek a theory,
not to propound one.'  (*The Noble Bachelor*)

(1) A key new ingredient in the study of functions of more than one variable is the idea of *determinacy*. A function $f: \mathbb{R}^n, 0 \to \mathbb{R}, 0$ is said to be $k$-determined if any function $g: \mathbb{R}^n, 0 \to \mathbb{R}, 0$ whose Taylor expansion at 0 coincides with that of $f$ up to and including terms of order $k$ is right equivalent to $f$. (We proved in chapter 3 that any function of one variable of type $A_k$ is $(k+1)$-determined.) Investigate this notion by reading chapter 11 of Bröcker and Lander (1975), chapter IV §3 of Gibson (1979) and pp. 262–270 of Wall (1971). (This latter article of Bochnak and Lojasiewicz is especially relevant. Theorem 1 there gives a weak version of Mather's criterion for $k$-determinacy for analytic functions. The construction of a 'formal' change of coordinates is simple and brings out the use of the conditions in the criterion rather nicely. One could prove that this 'change of coordinates' is analytic by the type of methods used in chapter 10, but the authors prefer to appeal to a difficult theorem of M. Artin.)

(2) Investigate the classification of functions of more than one variable by reading chapter IV §4 of Gibson (1979).

# APPENDIX

## *Null sets and Sard's Theorem*

'That is of enormous importance,' said Holmes,
making a note upon his shirt cuff. (*The Naval Treaty*)

We have used, principally in chapters 8 and 9, the concept of null set and Sard's fundamental theorem on critical values. In this appendix we shall define null sets, and prove a relatively trivial version of Sard's Theorem. This weak version nevertheless suffices for most of the applications we have given in this book. In particular it covers all the central applications in chapter 9, to generic geometry. This is, in fact, one attractive feature of the approach we use there. See Milnor (1965) or Gibson (1979) for complete proofs of Sard's Theorem.

**A1    Definition**
   (1) A *cube* $C$ in $\mathbb{R}^n$ of side $r$ is a product of $n$ open intervals $I_1 \times \cdots \times I_n$ each of length $r$ and parallel to the coordinate axes in $\mathbb{R}^n$. The *volume* of the cube $C$, denoted by vol. $C$ is $r^n$.
   (2) A set $L \subset \mathbb{R}^n$ is called a *null set in* $\mathbb{R}^n$ (or has measure zero) if, given any $\varepsilon > 0$, $L$ is contained in a countable union of cubes $C_i$, $1 \leqslant i < \infty$, with the sum of their volumes $\sum_{i=1}^{\infty}$ vol. $C_i < \varepsilon$.

One fundamental fact concerning null sets is the following.

**A2    Proposition**
   *If $L_j \subset \mathbb{R}^n$, $1 \leqslant j < \infty$ is a sequence of null sets in $\mathbb{R}^n$ then their union $L = \bigcup_{j=1}^{\infty} L_j$ is also a null set in $\mathbb{R}^n$.*

**Proof** Given $\varepsilon > 0$ we can cover each set $L_j$ by countably many cubes $C_{j,k}$ with $\sum_{k=1}^{\infty}$ vol. $(C_{j,k}) < \varepsilon/2^j$. Consequently the union $L$ is covered by countably many cubes $C_{j,k}$ with $\sum_{j=1}^{\infty} \sum_{k=1}^{\infty}$ vol. $(C_{j,k}) < \sum_{j=1}^{\infty} \varepsilon/2^j = \varepsilon$ as required. $\qquad\qquad\square$

The following property of null sets is the key to our version of Sard's theorem.

### A3 Proposition

*Let $U$ be an open subset of $\mathbb{R}^n$, $f: U \to \mathbb{R}^n$ a smooth mapping. If $L \subset U$ is a null set in $\mathbb{R}^n$ then $f(L)$ is a null set in $\mathbb{R}^n$.*

***Proof*** The set $U$ can be written as a countable union of cubes each of whose closures is contained in $U$. (Exercise A7(1).) So by A2 it is enough to show that for any such cube $C$ the set $f(C \cap L)$ is null. By the Mean Value Theorem (from advanced calculus) there is a constant $K$ such that for any $x$, $y$ in $C$

$$\|f(x) - f(y)\| < K\|x - y\|. \tag{1}$$

Suppose $\varepsilon > 0$ is given. Since $C \cap L$ is a null set it can be covered by countably many cubes $C_i$ of side $r_i$ with total volume $\sum_{i=1}^{\infty} r_i^n < \varepsilon$. Now the distance between any two points in a cube of side $r$ is $< n^{\frac{1}{2}}r$ (see exercise A7(2)). So using the inequality (1) above the distance between any two points of $f(C_i \cap L)$ is $\leqslant Kn^{\frac{1}{2}}r_i$, so that $f(C_i \cap L)$ is contained in a cube of side $2Kn^{\frac{1}{2}}r_i$ (see exercise A7(3)). Thus since $\sum_{i=1}^{\infty} (2Kn^{\frac{1}{2}}r_i)^n = (2Kn^{\frac{1}{2}})^n \sum_{i=1}^{\infty} r_i^n < (2Kn^{\frac{1}{2}})^n \varepsilon$, and $\varepsilon > 0$ was arbitrary, this shows that $f(C \cap L)$ is null. $\qquad \square$

We now generalize our definition of a null set.

### A4 Definition

Let $X \subset \mathbb{R}^N$ be a smooth $n$-manifold. A set $L \subset X$ is *a null set in $X$* if there is a countable collection of parametrizations $\gamma_i: U_i \to X$ with $\gamma_i(U_i)$ covering $X$ and $\gamma_i^{-1}(L)$ a null set in $\mathbb{R}^n$, for each $i$.

In fact any manifold $X$ has a countable collection of parametrizations which cover $X$, and using A3 it is not difficult to prove that our definition of null set is independent of the choice of such parametrizations. (See exercise A7(5).)

We are now in a position to prove Sard's Theorem.

### A5 Sard's Theorem (easy version)

*Let $X \subset \mathbb{R}^N$, $Y \subset \mathbb{R}^M$ be smooth manifolds, $f: X \to Y$ a smooth map, and suppose further that* dim $X <$ dim $Y$. *Then $f(X)$ is a null set in $Y$, and in particular the set of critical values of $f$ is null in $Y$.*

*Proof* *Step 1* Let dim $X=n$, dim $Y=m$. Claim: $X\times\{0\}$ is a null set in $X\times\mathbb{R}^{m-n}$. For if $\gamma\colon U\to X$ is a local parametrization of $X$, the map $(\gamma,\text{id})\colon U\times\mathbb{R}^{m-n}\to X\times\mathbb{R}^{m-n}$, $(\gamma,\text{id})(u,v)=(\gamma(u),v)$ is a parametrization of of $X\times\mathbb{R}^{m-n}$. But $(\gamma,\text{id})^{-1}(X\times\{0\})=U\times\{0\}$, which is clearly a null set in $\mathbb{R}^{n}\times\mathbb{R}^{m-n}$. (See exercise A7(4).)

*Step 2* Define $F\colon X\times\mathbb{R}^{m-n}\to Y$ by $F(x,a)=f(x)$. The result now follows from step 1 and the following:

*Claim* If $F\colon Z\to Y$ is a smooth mapping of smooth manifolds with dim $Z=$ dim $Y=m$, and $L\subset Z$ is a null set in $Z$, then $F(L)$ is a null set in $Y$. To prove this we first use definition A4 to obtain a countable collection of parametrizations $\gamma_i\colon W\to Z$ with $\gamma_i^{-1}(L)$ a null set in $\mathbb{R}^m$ for each $i$. We can find a countable collection of parametrizations $\delta\colon V\to Y$ covering $Y$. We want to show that $\delta^{-1}(F(L))$ is a null set in $\mathbb{R}^m$ for each $\delta$. Since the $\gamma_i(W)$ cover $Z$ it is clearly enough to show, in view of A2, that $\delta^{-1}(F\circ\gamma_i)(\gamma_i^{-1}(L))$ is a null set in $\mathbb{R}^m$ for each $i$ and $\delta$. Now $\gamma_i^{-1}(L)$ is a null set in $\mathbb{R}^m$, hence so is $\gamma_i^{-1}(L)\cap(F\circ\gamma_i)^{-1}(\delta(V))=L_1$. But $\delta^{-1}\circ F\circ\gamma_i\colon (F\circ\gamma_i)^{-1}(\delta(V))\to V$ is a smooth map between open subsets of $\mathbb{R}^m$; consequently by A3 $\delta^{-1}\circ F\circ\gamma_i(L_1)=\delta^{-1}\circ F\circ\gamma_i(\gamma_i^{-1}(L)\cap(F\circ\gamma_i)^{-1}(\delta(V)))$ is a null set in $\mathbb{R}^m$. However this latter set is easily seen to coincide with $\delta^{-1}((F\circ\gamma_i(\gamma_i^{-1}(L)))$ as required.    □

Please check that this version of Sard's Theorem suffices to prove Thom's Transversality Lemma 8.17 in the case when (in the notation of 8.17) $m-$ dim $Y<$ dim $X$. In other words we only require this easy version of Sard's Theorem in situations (such as in chapter 9) when $f\colon X\to\mathbb{R}^m$ transverse to $Y\subset\mathbb{R}^m$ actually means $f$ 'misses' $Y$.

Finally we establish the following result which is needed for the usual application of Thom's Lemma, that is, proving transverse maps are dense.

## A6    **Proposition**
*If $L\subset\mathbb{R}^n$ is a null set then its complement is dense in $\mathbb{R}^n$.*

*Proof* If $\mathbb{R}^n-L$ is not dense in $\mathbb{R}^n$ we can find a non-empty cube $C$ whose closure $cl(C)$ is a subset of $L$. Since $L$ is a null set, given any $\varepsilon>0$ there is a countable family of cubes $C_i$ with $L\subset\bigcup_{i=1}^{\infty}C_i$ and $\sum_{i=1}^{\infty}$ vol. $C_i<\varepsilon$. Choose $\varepsilon=$ vol. $C$. As $cl(C)\subset L$ is compact it has a finite cover $C_{i_1},\ldots,C_{i_r}$ and we have vol. $C\leqslant\sum_{j=1}^{r}$ vol. $C_{ij}\leqslant\sum_{i=1}^{\infty}$ vol. $C_i<$ vol. $C$: the required contradiction.    □

The following exercises fill in minor steps in the above proofs. Another exercise is the first inequality of the last sentence of the proof above. One proof of this can be found on p. 33 of Golubitsky and Guillemin (1973).

## A7     Exercises

(1) Show that any open set $U$ in $\mathbb{R}^n$ can be written as the countable union of cubes, each of whose closures is contained in $U$. (Consider all cubes with rational centres and rational sides which have this property.)

(2) Show that the distance between any two points in a cube of side $r$ in $\mathbb{R}^n$ is $< n^{\frac{1}{2}} r$.

(3) If $A \subset \mathbb{R}^n$ has the property that the distance between any two points of $A$ is less than $d$, show that $A$ is contained in a cube of side $2d$. (Actually one can improve this to $d$.)

(4) (*i*) Show that any cube $C \subset \mathbb{R}^n$ of side $r$ can be covered by $N^n$ cubes each of side $r/(N-1)$, for any $N \geqslant 2$.
    (*ii*) Deduce that, if $m \geqslant 1$, the set $C \times \{0\} \subset \mathbb{R}^n \times \mathbb{R}^m$ is a null set in $\mathbb{R}^{n+m}$.
    (*iii*) Now prove that $\mathbb{R}^n \times \{0\} \subset \mathbb{R}^n \times \mathbb{R}^m$ is a null set in $\mathbb{R}^{n+m}$.

(5) Show that any manifold $X \subset \mathbb{R}^N$ has a countable collection of charts which cover $X$. (For each point $x \in X \subset \mathbb{R}^N$ there are open neighbourhoods $U_x$ of 0, $V_x$ of $x$ in $\mathbb{R}^N$ and a diffeomorphism $\phi: U_x \to V_x$ with $\phi(U_x \cap \mathbb{R}^n \times \{0\}) = X \cap V_x$. What we do is to produce a covering by coordinate charts which are given by balls in $\mathbb{R}^n$ with rational radius and whose centre has rational coordinates. For every point of $X$ with rational coordinates choose such a ball contained in a coordinate chart $V_x$: this is the new coordinate chart for this point. When $x$ has an irrational coordinate choose a coordinate chart $V_x$ and choose a ball (*i*) inside $V_x$, (*ii*) having rational radius and centre, (*iii*) containing $x$. Clearly this is possible. Since the balls with rational radius and centre are countable, this completes the construction.)

# Further reading

'The authorities are excellent at amassing facts,
though they do not always use them to advantage.'
(*The Naval Treaty*)

A sobering thought indeed. We have done our best not simply to amass facts, but to use them to some advantage. Certainly the books we have chosen for our very short list of recommended further reading belie Holmes's dictum. Our list is short because long lists are unnecessarily daunting – and the reader can extend the list (almost) indefinitely by following up references within the references, and so on, back to Newton and beyond. We make pretence only at *in*completeness, and follow our personal tastes and preferences. Geometry and its applications form a vast and wonderful achievement, and no one man (not even two) can comprehend more than a small corner of it. We hope that you will find a corner to explore for yourself.

1 *More on functions* Bröcker and Lander (1975), Arnold (1981).
2 *Stable maps* Gibson (1979), Martinet (1982) (these two also concern themselves with functions), Golubitsky and Guillemin (1974).
3 *Differential geometry* Thorpe (1979), DoCarmo (1976), Banchoff, Gaffney and McCrory (1982).
4 *Catastrophe theory* Poston and Stewart (1978), Zeeman (1977).
5 *Differential topology* Milnor (1965), Guillemin and Pollack (1974).

# References

'With all these data you should be able to draw some
just inference.' (*The Sign of Four*)

R. P. Agnew (1960), *Differential Equations*, McGraw-Hill, New York.

L. V. Ahlfors (1966), *Complex Analysis* (2nd ed.), McGraw-Hill, New York.

V. I. Arnold (1976), Wavefront evolution and equivariant Morse lemma, *Comm. Pure Appl. Math.* 29, 557–82.

V. I. Arnold (1981), *Singularity Theory – Selected Papers*, London Math. Soc. Lecture Notes 53, Cambridge University Press.

T. Banchoff, T. Gaffney and C. McCrory (1982), *Cusps of Gauss Mappings*, Research Notes in Mathematics 55, Pitman, London.

Th. Bröcker and K. Jänich (1982), *Introduction to Differential Topology*, Cambridge University Press.

Th. Bröcker and L. Lander (1975), *Differentiable Germs and Catastrophes*, London Math. Soc. Lecture Notes 17, Cambridge University Press.

J. W. Bruce (1983), A Note on First Order Differential Equations of Degree Greater than One and Wavefront Evolution, *Bull. London Math. Soc.* (to appear)

J. W. Bruce, P. J. Giblin and C. G. Gibson (1981), On Caustics of Plane Curves, *American Math. Monthly*, **88**, 651–667.

J. W. Bruce, P. J. Giblin and C. G. Gibson (1982), On Caustics by Reflexion, *Topology*, **21**, 179–199.

M. DoCarmo (1976), *Differential Geometry of Curves and Surfaces*, Prentice-Hall, Englewood Cliffs, New Jersey.

H. S. M. Coxeter (1969), *Introduction to Geometry* (2nd ed.), John Wiley, New York.

L. P. Eisenhart (1909), *A Treatise on the Differential Geometry of Curves and Surfaces*, Ginn and Co., Boston.

C. G. Gibson (1979), *Singular Points of Smooth Mappings*, Research Notes in Math. 25, Pitman, London.

M. Golubitsky and V. Guillemin (1973), *Stable Mappings and their Singularities*, Graduate Texts in Math. 14, Springer-Verlag, Heidelberg.

M. Golubitsky and D. Schaeffer (1979), A Theory for Imperfect Bifurcation via Singularity Theory, *Comm. Pure Appl. Math.* **32**, 21–98.

E. Goursat (1917), *A Course in Mathematical Analysis*, Vol. 1, Ginn and Co., Boston.

V. Guillemin and A. Pollack (1974), *Differential Topology*, Prentice-Hall, Englewood Cliffs, New Jersey.

M. W. Hirsch (1976), *Differential Topology*, Graduate Texts in Math. 33, Springer-Verlag, Heidelberg.

A. Kas and M. Schlessinger (1972), On the Versal Deformation of a Complex Space with an Isolated Singularity, *Math. Ann.* **196**, 23–9.

E. J. N. Looijenga (1974), *Structural Stability of Smooth Families of $C^\infty$ Functions*, Thesis, University of Amsterdam.

J. Martinet (1982), *Singularities of Smooth Functions and Maps*, London Math. Soc. Lecture Notes 58, Cambridge University Press.

J. W. Milnor (1965), *Topology from the Differentiable Viewpoint*, The University Press of Virginia, Charlottesville.

I. R. Porteous (1971), The Normal Singularities of a Submanifold, *J. Diff. Geom.* **5**, 543–564.

T. Poston and I. N. Stewart (1976), *Taylor Expansions and Catastrophes*, Research Notes in Math. 7, Pitman, London.

T. Poston and I. N. Stewart (1978), *Catastrophe Theory and its Applications*, Surveys and Reference Works in Math. 2, Pitman, London.

T. Poston and A. E. R. Woodcock (1974), *A Geometrical Study of the Elementary Catastrophes*, Lecture Notes in Math. 373, Springer-Verlag, Heidelberg.

P. T. Saunders (1980), *An Introduction to Catastrophe Theory*, Cambridge University Press.

D. M. Y. Sommerville (1924), *Analytical Conics*, G. Bell, London.

M. Spivak (1965), *Calculus on Manifolds*, W. A. Benjamin, New York.

R. Thom (1975), *Structural Stability and Morphogenesis*, W. A. Benjamin, Reading, Massachusetts.

J. A. Thorpe (1979), *Elementary Topics in Differential Geometry*, Springer-Verlag, Heidelberg.

C. T. C. Wall (ed.) (1971), *Proceedings of Liverpool Singularities Symposium* I, Lecture Notes in Math. 192, Springer-Verlag, Heidelberg.

C. T. C. Wall (1976), *Geometric Properties of Generic Differentiable Manifolds*, *Geometry and Topology* III, Lecture Notes in Math. 597, 707–774, Springer-Verlag, Heidelberg.

T. J. Willmore (1959), *An Introduction to Differential Geometry*, Oxford University Press.

E. C. Zeeman (1977), *Catastrophe Theory, Selected Papers* 1972–77, Addison-Wesley, London.

J. Zeitlin (1981), Nesting Behavior of Osculating Circles and the Fresnel Integrals, *Math. Mag.* **54**, 76–80.

# Index of notation

'The object of those who invented the system has apparently been to conceal that these characters convey a message . . . .'

(*The Dancing Men*)

# Index

'In a morass, Watson?'
'I am at my wits' end.'
'Tut, tut; we have solved some worse problems.
At least we have plenty of material, if we can
only use it.'
*(The Priory School)*

*Note.* Boldface references are to
definitions and statements of theorems.
Italic references indicate the presence of a
diagram, but do not exclude text material.

Agnew, R. P. 100
Ahlfors, L. V. 191
almost all
  algebraic surfaces 167
  maps $\mathbb{R}^n \to \mathbb{R}^m$ 132
  maps are transverse 163
  matrices 165
  points regular values 118
  questions 121
  translations give transversality 165
  unfoldings are (p)versal 108, 167
analytic function(s) 45, 50, 103, 186*ff*
  example of non 45
  $k$-determinacy of 205
antiorthotomic **132**, *133*
  cusp on 132
apparent contour, see contour
applied mathematics 90
arc-length **25**
  of evolute 33
  of envelope 88
Arnold, V. I. 44, 129, 153, 200, 202, 205,
    210
Artin, M. 200, 205
$A$ singularity **43**, 47, 202 (see also distance-
    squared function, height function)
  $A_l$ close to $A_k$ 107
  bifurcation set of 110, 111, 113–15
  discriminant set of 117–18
  measuring contact 71
  near $t^5$ *102*
  only $A_1$ 62
  (p)versal unfolding 106, 107
  two variables 196
  unfolding of 103

versal unfolding of 116, 117
astroid **79**, *79*, 80, 88
Banchoff, T. 125, 205, 210
based vector **63**
Berry, M. 203
bifurcation set 5
  of unfolding **110**
  as critical values of projection 110
  uniqueness of 111
  of $A_k(k \leqslant 4)$ *113–15*
  of distance-squared function 122, 123
  of simple singularities *202*
binormal(s) **36**
  set of 124, *125*
  formula for 37
Bochnak, J. 205
Brieskorn, E. 200
Bröcker, Th. 41, 46, 56, 71, 103, 120, 130,
    171, 186, 205, 210
Bruce, J. W. 141, 142, 149, 153
bump function **46**
butterfly 98, 120
  catastrophe 202, 203
Cantor set 56
catastrophe
  elementary 202*ff*
  machine 1, 8, 203
  manifold 83
  surface 3, 70, 98
  theory (CT) 202*ff*
Cauchy's inequalities **190**
caustic by reflexion 9, *74*, 89, **141***ff*
  of circle *95*, 144*ff*, *146*, 148
  cusp on 143, 148
  of ellipse *146*, 148
  as evolute of orthotomic 94
  higher cusps on 149
  light source on normal 146–7
chain rule **66**, **157**
circle(s) 13
  caustic of *95*, 144*ff*, *146*

215

LIBRARY
OF
MOUNT ST. MARY'S
COLLEGE
EMMITSBURG. MARYLAND